AF553587

TEXTBOOK OF PHOTOBIOLOGY

TEXTBOOK OF PHOTOBIOLOGY

Dr. Shubhrata R. Mishra
Dept. of Botany
Vikram University
Ujjain (M.P.)
(India)

DISCOVERY PUBLISHING HOUSE PVT. LTD.
NEW DELHI-110 002

First Published-2010

ISBN 978-81-8356-569-1

Published by:

DISCOVERY PUBLISHING HOUSE PVT. LTD.
4831/24, Ansari Road, Prahlad Street
Darya Ganj, New Delhi-110002 (India)
Phone: +91-11-23279245, 43764432 • Fax: +91-11-23253475
E-mail: parul.wasan@gmail.com
info@discoverypublishinggroup.com
web: www.discoverypublishinggroup.com

Printed at:
Arora Offset Press, Delhi – 92

Preface

Sunlight is one of the most important elements in our environment. Plants harvest the energy of sunlight in order to grow, and thus to provide food for other organisms. Sunlight also provides information needed by organisms to trigger numerous biological responses. These are the beneficial wavelengths of sunlight.

Sunlight also has its bad side. The shorter wavelengths (or the longer wavelengths plus photosensitizers) can kill plants and other organisms, and produce cancer and other debilitating conditions in man.

Photobiology is the study of both the good and the bad effects of light. Studies range from the atomic level to the level of communities of organisms. Photobiologists use all of the tools of science to study the chemical and biological effects of light. Photobiology is an exciting and challenging field of science.

Light is composed of photons, and is propagated in the form of waves. Wavelength is the distance in the line of advance of a light wave from the bottom of one trough to the bottom of the next trough. The photons at each wavelength have different energies; the shorter the wavelength (nm) the higher the energy.

The shortest wavelength of sunlight reaching the surface of the earth is at about 295 nm. Wavelengths shorter than this are filtered out by the stratospheric ozone layer. If the ozone layer is attenuated, then more short wavelength UV radiation will reach the earth, and will have a profound deleterious effect on man, animals, plants, and other organisms. The sun has both detrimental and beneficial effects on organisms, and all organisms have developed defence strategies (behavioural and biochemical) to protect themselves from the harmful wavelengths and harmful intensities, while optimizing the receipt of the beneficial wavelengths and intensities.

In general, the term "light" is used to define those wavelengths that are visible to man, although other organisms can "see" in the UV region. However, it is not uncommon for some people to use the term "UV light". UV radiation is the preferred term.

The spectrum of sunlight on earth during a typical day. Since life on earth evolved under the sun, whose terrestrial photon flux ("intensity") is greatest between 400 nm and 800 nm, it should not be surprising that most biological responses to light are induced by radiation between 400 nm and 800 nm. Very little radiation below 300 nm reaches the surface of the earth, because of its absorption by stratospheric ozone; very little radiation above 1000 nm reaches the surface of the earth because emission from the sun is low in this region, and because atmospheric water absorbs strongly above 1000 nm.

– Author

Contents

1

Introduction

"Photobiology is broadly defined to include all biological phenomena involving non-ionizing radiation. It is recognized that photobiological responses are the result of chemical and/or physical changes induced in biological systems by non-ionizing radiation."

Non-ionizing radiation produces excited states in molecules due to the absorption of one or more photons. Excited-state molecules can react with adjacent molecules, but most frequently they undergo photochemical and photophysical changes within their own molecular structure. Non-ionizing radiation is grouped into three main regions; Ultraviolet (UV) radiation (short wavelengths that are not visible to man), Visible radiation (longer wavelengths than UV radiation), and Infrared radiation (still longer wavelengths, and also not visible to man).

The UV region is generally divided into three regions (especially in Photomedicine), i.e., the UV-C region, which is generally defined as being in the wavelength region from 100-280 nanometers (nm), the UV-B region as 280-320 nm, and the UV-A region as 320-400 nm. Other terms commonly used for the UV region are Far-UV (210-300 nm) and Near-UV (300-380 nm).

The Visible region is generally defined as 400-760 nm, and the Infrared region lies above 760 nm.

Light is composed of photons, and is propagated in the form of waves. Wavelength is the distance in the line of advance of a light wave from the bottom of one trough to the bottom of the next trough. The photons at each wavelength have different energies; the shorter the wavelength (nm) the higher the energy.

The shortest wavelength of sunlight reaching the surface of the earth is at about 295 nm. Wavelengths shorter than this are filtered out by the stratospheric ozone layer. If the ozone layer is attenuated, then more short wavelength UV radiation will reach the earth, and will have a profound deleterious effect on man, animals, plants, and other organisms. The sun has both detrimental and beneficial effects on organisms, and all organisms have developed defense strategies (behavioral and biochemical) to protect themselves from the harmful wavelengths and harmful intensities, while optimizing the receipt of the beneficial wavelengths and intensities.

In general, the term "light" is used to define those wavelengths that are visible to man, although other organisms can "see" in the UV region. However, it is not uncommon for some people to use the term "UV light". UV radiation is the preferred term.

The spectrum of sunlight on earth during a typical day. Since life on earth evolved under the sun, whose terrestrial photon flux ("intensity") is greatest between 400 nm and 800 nm, it should not be surprising that most biological responses to light are induced by radiation between 400 nm and 800 nm. Very little radiation below 300 nm reaches the surface of the earth, because of its absorption by stratospheric ozone; very little radiation above 1000 nm reaches the surface of the earth because emission from the sun is low in this region, and because atmospheric water absorbs strongly above 1000 nm.

Photobiology is a large discipline that includes studies of both the beneficial and harmful effects of light. It covers topics from the atomic level to that of ecological communities. Photobiologists use all of the tools of science to study the chemical and biological effects of light and other non-ionizing radiation.

THE DIFFERENT SPECIALITY AREAS OF PHOTOBIOLOGY

Photobiology can be divided into 12 major speciality areas. Eleven of these are concerned with the absorption of light in a biological system, and one is concerned with the emission of light by biological systems (Bioluminescence). These areas are briefly defined below, and will be more fully described in appropriate modules.

1. **Photophysics.** This specialty area is concerned with the physical interactions of light with matter at the atomic and molecular level. These include the vibration and rotation of molecules.

2. **Photochemistry.** This is the study of the chemical changes that occur in molecules after the direct absorption of light energy (compare with Photosensitization). These include both alterations in the absorbing molecule, and reactions that occur between the absorbing molecule in its excited state and an adjacent molecule.

   The First Law of Photochemistry states that "Light must be absorbed before photochemistry can occur". The power of this law is that by knowing the absorption spectrum of a molecule, i.e., by knowing which wavelengths of light can be absorbed by a molecule, one can immediately predict what wavelengths of light can have a photochemical effect on that molecule, and also what wavelengths of light will have no effect (since they are not absorbed).

3. **Photosensitization.** In this process, the light energy is absorbed by one type of molecule (the sensitizer), and the resulting energy-rich state(s) of the sensitizer then undergoes reactions that ultimately result in the chemical alteration of another type of molecule in the system (the substrate molecule). The sensitizer is not altered in certain types of photosensitization reactions.

   Nearly all organisms contain molecules that are potential photosensitizers (e.g., bilirubin, chlorophylls, and porphyrins). Photodynamic therapy, which employs a photosensitizing drug and light, has important applications in cancer therapy. Certain forms of lung cancer and esophageal tumors are treated by photodynamic therapy to target and destroy cancer cells.

4. **UV Radiation Effects on Molecules and Cells.** This field is concerned with the UV radiation photochemistry of deoxyribonucleic acid (DNA), ribonucleic acid (RNA) and proteins, and the biological effects produced by the photochemical and photophysical changes in these molecules (e.g., lethality, mutations). The field is also concerned with the sophisticated biochemical systems by which the cells can repair this photochemical damage.

5. **Environmental Photobiology.** The different wavelengths of sunlight exert both beneficial and detrimental effects, not only on individual cells and organisms, but more importantly on the whole ecosystem, where one deals with the effects of light on species composition and productivity.

6. **Photomedicine.** This field is concerned both with the detrimental effects and the beneficial effects of non-ionizing radiation. In Photomedicine, one most often thinks in terms of sunlight-induced skin cancer, but there are many other important topics. Then there is the beneficial area, where light alone, or sensitizers plus light are used to treat certain clinical conditions, e.g., psoriasis and cancer.

   Photomedicine also includes the field of Photoimmunology, e.g., the absorption of light can modulate the immune system of the body, and thus prevent the immunological rejection of tumors.

7. **Non-Ocular Photoreception.** As the term implies, light is received by a receptor in an organism, other than the eye, to monitor the environment. A few examples are the circadian clock, which controls hormonal levels in birds and animals, and photoperiodism, which controls the seasonal growth and subsequent shrinking of reproductive organs, e.g., in birds.

8. **Vision.** The photoreception that occurs through the eye is called Vision. This field deals with the structure and photochemistry of the visual pigments, and the structure of the receptor cells.

9. **Photomorphogenesis.** The development of an organism can be influenced by the information in light. This information comes from the quantity, the quality (i.e., wavelengths present), the spatial asymmetry (i.e., the direction from which the light comes), and the periodicity of the light. Some examples of photomorphogenesis are the germination of light sensitive seed, and the flowering of long-day plants.

10. **Photomovement.** To produce movement, plants and organisms depend upon the quality and the direction of the light striking their photoreceptors. In Photokinesis, an organism swims toward or away from light. Phototropic curvature in plants can occur toward or away from the light. Perhaps the best know example of this is sunflowers.

    Charles Darwin, best known for developing the theory of evolution by natural selection, collaborated with his son Francis in writing an early book on phototropism, *The Power of Movement in Plants* (1880). This book has been so influential that Darwin would be well-known to biologists even if he had not written his great books on evolution.

11. **Photosynthesis.** It is not the information in the light that is used in photosynthesis, rather it is the energy of the light that is converted to stabilized chemical energy. This involves the absorption of light by a pigment, energy transfer, energy trapping or stabilization by reaction centers, and the initiation of chemical reactions from donor to acceptor molecules. This is a light harvesting reaction, while most of the other photobiological reactions require only a few photons to trigger responses.

12. **Bioluminescence.** For most people, bioluminescence is represented by the flash of a firefly or the phosphorescence that frequently occurs on agitating the surface of the ocean. Bioluminescence is the highly efficient cold-light emission that has a biological function for the organism concerned, e.g., finding a mate or food. More than half of all phyla in the animal kingdom contain members that are bioluminescent.

In *Nature's Light*, Francine Jacobs relates a story in which the bioluminescent fire beetle (*Pyrophorus noctilucus*) may have changed the history of the Americas. "In 1634, when the English were about to land at night on the island of Cuba, they saw many lights. Mistakenly, they believed them to be torches held by Spanish forces already on the island. Deciding that they were greatly outnumbered, the English withdrew and sailed on. What they probably observed were the glowing lights of fire beetles."

Sunlight is one of the most important elements in our environment. Plants harvest the energy of sunlight in order to grow, and thus to provide food for other organisms. Sunlight also provides information needed by organisms to trigger numerous biological responses. These are the beneficial wavelengths of sunlight.

Sunlight also has its bad side. The shorter wavelengths (or the longer wavelengths plus photosensitizers) can kill plants and other organisms, and produce cancer and other debilitating conditions in man.

Photobiology is the study of both the good and the bad effects of light. Studies range from the atomic level to the level of communities of organisms. Photobiologists use all of the tools of science to study the chemical and biological effects of light. Photobiology is an exciting and challenging field of science.

Sunlight provides:

- Energy heat and photosynthesis
- Information photomorphogenesis; photoperiodism, circadian rhythms

What is the nature of light?

How does light interact w/plants (i.e. which biochemicals "sense" light & how are they affected?

Visible radiation = light

One type of electromagnetic energy, » = 400 700 nm

UV > visible > infrared = 100 nm 800+ nm

Plants use light of 400 700 nm [capable of exciting molecules without destroying them (breaking bonds)]

Light's physical nature is both waves and particles

As a wave (a continuous wave) when light is transmitted:

reflected

refracted

As a particle when light is:

absorbed by pigments

emitted as a photon (e.g. fluorescence or phosphorescence)

Light (electromagnetic energy) as continuous waves

A wave = a regular continuous change

Electro-magnetic energy (or Electro-magnetic Radiation)

Electro (an electrical wave) + Magnetic (A magnetic wave) vectors that oscillate at 90° to each other

Wavelengths (») are expressed in nm in biology, we (biologists) usually use » (notn)

Frequency = # of wave crests per second

$C = 3 \times 10^8 ms^{-1}$ (speed of light)

n = frequency

Light as a particle = a photon

Each photon has a certain amount of energy, called a quantum [(expressed in Joules (J)]

This energy is related to the » and the n of the light

- Eq = hc/ n = hn
- h = Planck's constant = 6.62×10^{-34} Js/photon

Therefore, quantum energy of light is:

- **inversely** proportional to »
- **directly** proportional to n
- So . . . hn > represents a photon

To calculate the energy of a photon:

If » = 660 nm (= 6.6×10^{-7} m)

Eq = (6.62×10^{-34}Js photon-1)($3 \times 10^{8}ms^{-1}$)/(6.6×10^{-7}m)

Eq = 3.01×10^{-19} J/photon (of red light)

If » = 4.35×10^{-7} m(435 nm)

Eq = 4.56×10^{-19} J/photon (of blue light)

Usually the energy is expressed in moles...

...so multiply by 6.023 x 1023 photons/mole

for light of 660 nm (red light) 181 KJ/mole

for light of 435 nm (blue light) 274 KJ/mole

How many moles of photons are needed to excite a mole of pigment molecules in PSN?

The absorption of light energy by pigments (molecules that absorb light) (occurs in less than femtosecond (10^{-15} sec)

Two Principles in Photobiology are:

The "Gotthaus-Draper Principle": only absorbed light is active photochemically or photobiologically

The Einstein-Stark Law (photochemical equivalence): A singe photon can excite only one electron. (i.e. quanta cannot be subdivided and e-s cannot be partially excited)

Which pigments are useful to plants? they absorb light in a femptosecond (10-15 sec)

e.g. chlorophylls a and b, carotenoids, anthocyanins, phytochrome, phycobilins.

Pigments are necessary for all photobiological responses

Energy of the absorbed photon is transferred to an electron

Pigment in excited state = singlet state

Photons (light particles) e^-

Pigment (ground state = unexcited)

This creates a change in the energy level and electrons can only exist as discrete energy levels (as do photons)

Therefore, electrons cannot be partially excited.

Either the photon has the quantum energy to excite an electron or it doesn't!

A single photon excites only a single electron,

...but, pigments have many e-'s of different energy, -

hence, need photons of different »'s

Therefore chlorophyll, e.g. when exposed to white light (all » of visible light) has many excited energy levels at the same time.

Lifetime of an "excited" molecule?

One nanosecond = 10^{-9} second, then returns to ground state

Any excess energy must be dissipated How?

Thermal deactivation (heat is released)

fluorescence: emission of light of a lower energy content (usually red light is emitted, regardless of the original »)

inductive resonance (AKA radiationless transfer) e.g. transfer of energy from carotenoids to chlorophyll.

dropping from singlet to the "triplet" state (10-3 sec) (a metastable state, much more stable than singlet, so chemical reactions still occur)

Photo-oxidation of chlorophyll a and reduction of the acceptor and emdash:

- In PS II: A quinone
- In PS I: Ferredoxin

Absorption Spectrum and Action Spectrum

Absorption Spectrum (fingerprint) and emdash; efficiency (according to ») of absorption by any particular pigment (for example in PSN and emdash; chlorophyll a, chlorophyll b, phycocyanin)

Action spectrum and emdash; absorbance of light (»s) in a particular process (e.g. photosynthesis or photomorphogenesis)

Assumption is that light most efficiently absorbed is most effective in process (e.g. chlorophyll a absorbs most efficiently in red and blue » and most effective for PSN)

Comparison of these 2 types of spectra helps ID the pigment responsible for a photobiological process.

Measuring light &emdash; must be accurate

3 parameters

- Light quality = » (which ones?)
- Light quantity = how much light? (intensity)
- Timing = duration (periodicity)

Photon fluence - # of photons measured on a fluence detector (spherical)

Energy fluence &emdash; amount of energy measured (on the fluence detector)

Therefore, photon fluence rate = mol (of photons)/m^2s

Energy fluence rate = Joules/m^2s OR watts/m^2

Fluence vs. Irradiance?

Fluence and emdash; Quantity of (light) energy intercepted by a small sphere

Irradiance and emdash; Quantity of (light) energy intercepted by a flat surface

Most useful (in terms of photobiology) and emdash; fluence detectors

The best to use &emdash; those that only measure PAR (PSN active radiation), 400 nm and emdash; 700 nm.

Why isn't one made that measures 700 and emdash; 750 nm?

Far-Red fluence promotes phototropic response and phytochrome absorbs "light" in » > 700 nm.

Lux/footcandle and emdash; Measure of luminosity according to the human eye at »550 nm

Remember &emdash; old and outdated measurements are NOT accurate for plant physiology experiments.

Intensity vs. Dose?

Intensity and emdash; quantity of radiation (light) emitted from a source (such as a light bulb)

Dose and emdash; The energy from the light that is actually absorbed (not easily measured)

Light environment (radiation environment) of plants continually changes throughout the day and the seasons (fluence rate and wavelength)

Greenhouse Effect &emdash; absorption of infrared radiation by gases in the atmosphere

Ozone layer (O_3) and emdash; absorption of UV-C and UV-B radiation

Depletion of ozone > more UV-B radiation reaches earth's surface (higher energy than UV-A). Luckily, ozone absorbs UV-C better!

Photoreceptors and emdash; pigments that absorb light (radiation) for use in plant's physiological processes

Pigment molecules and emdash;

absorb radiant energy

process the energy

process information carried by the »s

Chlorophylls (Main PSN pigments)

Porphyrin head (a cyclic tetrapyrrole) of 4 pyrrole rings w/ a prosthetic Mg ion

Phytol (20 carbon alcohol) tail esterified to ring IV

Mg^{++} gives chlorophyll its green colour

- Mg > non-green Pheophytin

Phytol tail (an isoprene derivative) is lipid-soluble, anchors chlorophyll a in thylakoids of chloroplasts to hydrophobic proteins

Four chlorophyll types are known and emdash:

- Chlorophyll a: primary psn pigment
- Chlorophyll b: -CHO substitutes for and emdash; CH_3 on ring II of the porphyrin head of chlorophyll
- Chlorophyll c: lacks the phytol tail of chlorophyll
- Chlorophyll d: -O-CHO substitutes for the and emdash; $CH=CH_2$ on ring I of the porphyrin head of chlorophyll

Protochlorophyll a and emdash; immediate precursor to chlorophyll a (accumulates in dark-grown seedlings)

Only difference: C=C between C7 and C8 in ring IV instead of a C and emdash; C for chlorophyll.

All chlorophylls absorb at distinctly different »s, though absorption spectra are similar

Terpenoids: All share a common synthetic pathway

Hormones (GA & ABA)

Carotenoid Pigments

Carotenes (no oxygen)

Xanthophylls (contain oxygen)

sterols (e.g. cardiac glycosides)

latex (natural rubber precursors)

Essential oils (mints, perfumes, etc.)

side chains of cytokinins & chlorophyll

pyrethroids (insect toxins, from chrysanthemums)

Are all polymers of isoprene units (5-carbon iso-pentane)

Carotenoids: orange and yellow pigments, accessory to psn C_{40} terpenoids (derived from the isoprenoid pathway)

Lipid soluble and emdash; predominantly hydrocarbon located in thylakoid membranes or in chromoplasts (can crystallize)

B-carotene and alpha-carotene (orange-reddish)

Lycopene

Xanthophylls (zeaxanthin, lutein) &emdash; yellow

Function of Carotenoids?

absorbs blue »s very strongly

passes the energy to chl a (?)

protect chl a and chl b from photo oxidation by reversibly combining w/O radicles to form xanthophylls

Phycobilins

- not in higher plants (except phytochromobilin), red algae, cyanobacteria
- H_2O soluble
- Accessory pigments
- Similar to bile pigments in livers of mammals & emdash; note how red livers are!)

Four known phycobilins (chromoproteins) accessory psn pigments (in phycobilisomes) and emdash:

phycoerythrin

phycocoyanin

allophycocyanin

phytochromobilin and emdash; regulates almost every stage of plant development i.e. phytochrome

chromoprotein and emdash; protein part = apoprotein

chromophore = absorbing part

chromophore + apoprotein = holochrome

"cryptochrome" and emdash; a potential pigment

What is responsive to Blue »'s & UV-A »'s and important for development in ferns, mosses, fungi? Phototropism

Action Spectra [for many developmental responses in lower plants & phototropic responses in high plants]

Two peaks &emdash:

At »'s of 400 nm and emdash; 475 nm and

At »'s of 320 nm and emdash; 400 nm

Not known which pigment absorbs and processes the information and emdash:

- Carotene and emdash; does absorb blue light
- Flavins and emdash; riboflavin, FMN (flavin mononucleotide), FAD (flavin adenine dinucleotide), can occur as flavoproteins

These seem to be the elusive cryptochromes, but not clearly established thus far.

UV-B Receptors

Action Spectra peaks at 290 nm = UV-B

Anthocyanin synthesis in sorghum vulgave

Anthocyanin synthesis in carrot & Parsley

(1% decrease in ozone leads to 2% increase in UV)

photoreceptor for UV-B 1

Flavonoids

Pigments of bracts, leaves, flowers, fruits

e.g.

anthocyanins and anthocyanidins (scarlet, pink, purple, blue)

chalcones & aurones (yellow flowers)

flavones (white flower petals)

phenylpropane derivatives $C_6–C_3–C_6$

Used as dyes for thousands of years, chemistry of flavonoids is well known in anthocyanidins and emdash; OH increases blueness

–OCH_3 increases redness

H_2O soluble

Primarily in plant's vacuoles (cell sap)

Colours of anthocyanins are sensitive to pH

Absorb strongly in »'s of 475 nm &emdash; 560 nm and also absorb in the UV-B »'s

(could be protecting leaf cells from UV-B damage!)

Physiologically, role(s) of flavonoids in plants is still unknown (more research needs to be done)

Isoflavonoids and emdash; plant pigments too

Antimicrobial (kin bact and fungi) one group of the phytoalexins

Induced to synthesize by presence of pathogens

Their polysaccharides, glycoproteins, proteins and emdash; these molecules are elicitors for production of phytoalexins (by the plants)

Depending on the plant family and emdash;

Elicitors of mRNA for plant enzymes involved in isoflavonoid synthesis (i.e. phytoalexins)

Elicitors of mRNA for plant enzymes involved in synthesis of other phytoalexin chemicals

Ques and emdash; difference between anthocyanins and betacyanins?

Phytochrome and emdash; photoreceptor for light photoperiod

two REVERSIBLE states PR PFR

(ratio) PR: PFR signals sunrise and sunset

Note: actual night length is measured by the biological clock

Ration PR: PFR also measures light quality, e.g. triggers elongation of shaded shoots

2

Basic Ultraviolet Radiation Photobiology

INTRODUCTION

Ultraviolet (UV) radiation is generally divided into three regions, especially in Photomedicine, i.e., UV-C (100-280 nm), UV-B (280-320 nm), and UV-A (320-400 nm). Virtually no UV-C radiation reaches the earth, because of strong filtering by the ozone layer, however, most experiments on the effects of UV radiation on cells are performed using UV-C (e.g., 254 nm), because of it efficiency in producing damage to cells, especially to their deoxyribonucleic acid (DNA).

While research on UV radiation is largely focused on its detrimental effects, UV radiation also has a very important beneficial effect, i.e., the production of Vitamin D in the skin of humans. Vitamin D3 is produced photochemically in the skin from 7-dehydrocholesterol. The highest concentrations of 7-dehydrochoiesterol are found in the epidermal layer of skin. Synthesis in the skin involves UV-B radiation (280-320 nm), which effectively penetrates only the epidermal layers of skin. While 7-dehydrocholesterol absorbs UV radiation at wavelengths between 270-300 nm, optimal synthesis occurs in a narrow band between 295-300 nm his module, however, will focus on the detrimental effects of UV radiation, and the many mechanisms that cells have for repairing this damage. These same repair systems are also very important for repairing metabolic damage to DNA, i.e., produced just from normal living and breathing.

Lethal and Mutagenic Effects of UV Radiation

UV radiation produces many types of photochemical alterations in DNA, ribonucleic acid (RNA), and protein, as well as to structures such as membranes. DNA, however, is the major target for the deleterious effects of UV radiation because it is the largest molecule in the cell, it is present in the fewest copies, it carries the genetic information for a cell, and it absorbs UV radiation very efficiently. In contrast, RNA and proteins are present in multiple copies, so that it is much more difficult to inactivate a significant number of these molecules in a cell by UV irradiation.

A large number of different types of damage are produced in DNA by UV irradiation. These include the modification of individual purine and pyrimidine bases (e.g., deamination, ring cleavage), the production of pyrimidine dimers (i.e., the covalent linkage of adjacent pyrimidines in DNA), and the addition of other molecules to the purines and pyrimidines (e.g., water (photohydrate), DNA-protein cross-linking).

The direct action of UV radiation does not produce strand breaks in DNA (but X-rays do), but single and double-strand breaks in DNA are produced as a by-product of DNA repair.

With photosensitized reactions, near-UV radiation or visible light activates a sensitizer; the sensitizer can then combine with DNA (e.g., psoralen, as in the treatment of psoriasis), or can transfer its energy to a new molecule. These reactions are highlighted in the modules on Photosensitization.

While the most drastic effect of UV radiation on a cell is lethality, UV radiation-induced damage can also result in mutations, i.e., alterations in the DNA base sequence of a gene such that the capacity of the gene product to perform an essential cellular function is altered. Some mutations can be lethal.

Fortunately, a cell has an enormous capacity to repair all types of damage to its DNA. These repair systems have names like photoreactivation, base excision repair, nucleotide excision repair, recombinational repair, and double-strand break repair.

These DNA repair systems have not evolved just to repair damage produced by UV radiation, they are also required to maintain the integrity of the DNA after damage by the products of normal metabolism (e.g., superoxide), and to repair the heat damage produced at body temperature (e.g., deamination). Some of these repair systems are very accurate, and do not produce mutations. Other systems, however, are efficient, but inaccurate, and do produce mutations.

The first indication that cells could recover from UV radiation-induced damage was the observation by Roberts and Aldous in 1949 that minor modifications in the postirradiation treatment of certain bacterial strains, resulted in a marked increase in their viability. Subsequently, it was observed that UV irradiated *E. coli* rec- cells showed enhanced survival when held under non-growth conditions for various periods of time before plating on growth medium to assay for survival. This process is called "liquid holding recovery". Thus, the delay of macromolecular synthesis (i.e., DNA, RNA, and protein) for a time after irradiation led to a higher survival. By holding cells in non-growth medium, it allows the cells to complete constitutive repair processes, e.g., nucleotide excision repair, without interference with the inducible recombinational repair systems in these repair-impaired rec strains.

The ultimate proof that cells could recover from radiation damage was the isolation in 1958 of a radiation sensitive mutant of *Escherichia coli* (Bs-1) by Ruth Hill. This discovery inaugurated the era of the genetic approach to the study of DNA repair and mutagenesis.

Genetic studies have been responsible for most of the discoveries about the biochemical basis of DNA repair and mutagenesis. For example, one can measure the initial yield of a particular type of radiation-induced lesion in the DNA of a wild-type bacterial strain (i.e., radiation resistant) and in a mutant strain (i.e., radiation sensitive), and then follow the kinetics and extent of the repair of that lesion, and thereby determine if the mutation under investigation has an effect on the repair of that type of DNA lesion.

Then by constructing strains that carry two mutations that affect DNA repair, one can determine through survival and DNA repair studies whether the two mutations act in the same or in different biochemical pathways for repair. For example, since the *uvrA recA* double mutant of *E. coli* is much more sensitive than either of the singly mutant strains, it indicates that the *uvrA* and *recA* genes function in separate pathways of DNA repair (Fig. 2.1).

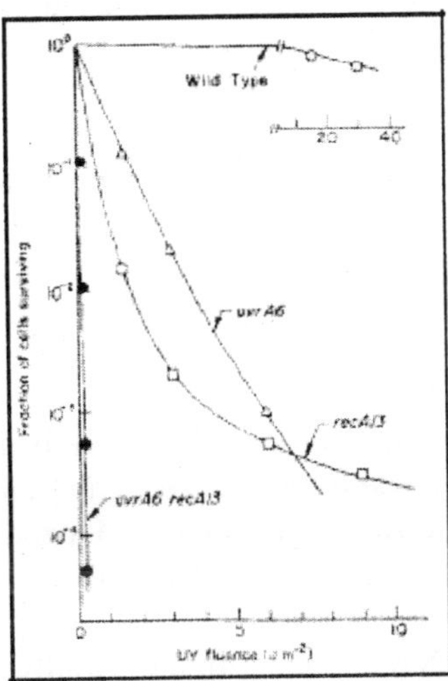

Fig. 2.1: UV radiation survival curves for DNA repair deficient mutants of *E. coli* K-12. The *uvrA6* mutation blocks nucleotide excision repair, and the *recA13* mutation blocks recombinational DNA repair. Note that the double mutant, *uvrA6 recA13* is very much more sensitive to UV radiation than either of the single mutants, indicating that the two single mutants are involved in separate pathways of DNA repair. Since the two single mutants have about the same sensitivity, it indicates that nucleotide excision repair and recombinational DNA repair are of about equal importance to the survival of UV irradiated *E. coli*.

The *uvrA* mutation blocks nucleotide excision repair, a repair system that is largely constitutive and error free. The *recA* mutation blocks recombinational repair, which is in part inducible by radiation damage, and therefore requires protein synthesis for it to work. This type of repair has a component that is error prone, i.e., it is efficient but makes mistakes, and thus leads to mutations. These two DNA repair systems are sometimes called "dark repair" systems, since they do not need the presence of light for them to function.

PHOTOREACTIVATION

There is a type of repair that requires light to function, and is found in a wide range of organisms. Thus, if *E. coli* are exposed to a lethal dose of UV radiation (e.g., 254 nm), lethality can be reduced or eliminated by a second exposure to light in the range of 300-450 nm. This process requires an enzyme, DNA photolyase (a flavoprotein), which binds to pyrimidine dimers. When this enzyme-substrate complex absorbs light in its flavine moiety at the above wavelengths, the pyrimidine dimer is split yielding undamaged pyrimidines, and the enzyme is released.

Nucleotide Excision Repair

In this process (Fig. 2.2), a distortion in the DNA caused by a lesion (e.g., pyrimidine dimer) is recognized by the UvrABC nuclease. It cuts on both sides of the lesion, producing a gapped structure that is about 12 nucleotides long. Repair replication, using DNA polymerase I, fills in the gap using the undamaged opposite strand of DNA as the template. Finally the break in the repaired DNA strand is sealed by DNA ligase (Petit and Sancar, 1999). This repair system is largely constitutive, and is error free.

Base Excision Repair

Altered purine and pyrimidine bases (e.g., thymine glycol, 3-methyladenine, uracil, urea) are recognized by a class of enzymes called DNA glycosylases, which are specific to varying degrees for particular types of altered bases. These enzymes split off the altered base at the N-glycosylic bond that links the base to the sugar in the DNA backbone. This site of the base loss is called an AP site (for apurinic or apyrimidinic). The AP site is degraded to produce a gap in the DNA strand, which is repaired by the same processes used to repair the gaps during nucleotide excision repair. This process is especially important for cells exposed to ionizing radiation and alkylating agents, but is also important for the repair of normal metabolic damage, including depurination at normal body temperature.

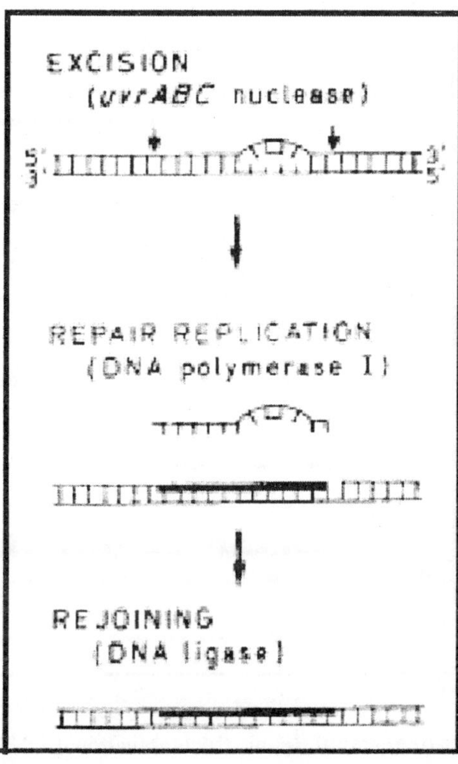

Fig. 2.2 Nucleotide Excision Repair in *E.coli*. The UvrABC nuclease recognizes the distortion in the DNA due to the lesion, then cuts on both sides of the lesion, producing a gapped structure about 12 nucleotides long. Repair replication (heavy line) fills the gap using the strand opposite the gap as the template. Finally, the break in the strand is sealed by DNA ligase.

Recombinational Repair

This type of repair is much more complicated than is excision repair, and requires many more gene products. The products of a number of these repair genes are induced by radiation damage, and therefore this type of repair requires protein synthesis before it can function. Because of its complexity, this type of repair makes mistakes (Fig. 2.3).

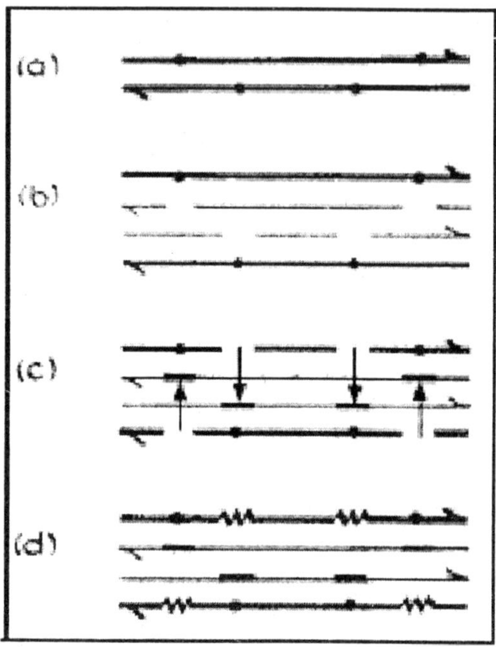

Fig. 2.3: Postreplication Repair (recombinational DNA repair) in UV irradiated *E. coli*. (a) The dots indicate lesions in the DNA. (b) DNA synthesis proceeds up to a lesion and then skips past the lesion, leaving a gap in the daughter strand. (c) Filling of the daughter strand gaps with DNA from parental strands by a recombinational process that requires a functional *recA* gene. (d) Gaps in the parental strands are repaired by repair replication (see Nucleotide Excision Repair).

Recombinational repair requires four strands of DNA (i.e., two sister duplexes). One type of recombinational repair is called Postreplication Repair, and comes into play after replication proceeds past a lesion in DNA, leaving a gap in the replicated strand opposite the lesion. This gap is then filled with material with the correct base sequence from the sister duplex strand. During this process of strand exchange from one sister duplex to another, however, slippage in the base pairing of these DNA strands can occur, leading to mutations, which can be lethal.

Recombinational Excision Repair

This process occurs in the part of the chromosome that was replicated prior to irradiation, i.e., where two sister duplexes were present before irradiation. After the excision of the lesion, the resulting gap is filled by the same recombinational process just described for postreplication repair (Fig. 2.4).

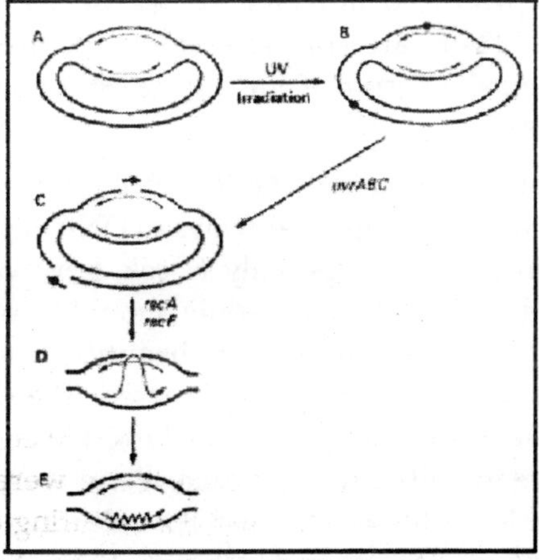

Fig. 2.4: The recombinational repair of excision gaps in *E. coli*. UV radiation-induced lesions are produced in both the replicated and unreplicated portions of the genome (B). The gaps produced by excision in the unreplicated portion are repaired by the classical methods of nucleotide excision repair (C). The gaps produced in the replicated portion of the chromosome are repaired by a recombinational process that requires both *recA* and *recF* (C-D).

Each of these "dark" repair systems involve the cutting of DNA strands, and this can lead to the formation of DNA double-strand breaks, which are lethal if not repaired. Fortunately, cells are very good at repairing DNA double-strand breaks by recombinational processes of the type just described, i.e., requiring the presence of two sister duplexes. These processes are called homologous recombination, since "proper" base pairing is involved.

Cells can also have the capacity for non-homologous recombination, where the ends of a double-strand break are simply sealed together, which usually results in the loss of bases, and therefore results in lethality or mutation.

Mutations are not only produced by errors in repair, they can also be produced by errors made during normal DNA synthesis. This was first suggested by the finding that some mutations are produced independent of the recombinational repair systems. There are special systems for the prevention, detection, and repair of errors made during DNA replication [e.g., Mismatch Repair].

Most of our early knowledge about the biochemistry and genetic control of the multiple pathways of DNA repair has come from studies on bacteria, especially *E. coli*, however, when the granting agencies decreed that people should work on human cells in order to obtain funding, the beautiful and not nearly complete work on bacteria essentially ceased. Then there were a few years of slow progress as people switched to eucaryotic cells (yeast and human cells). At that time, there were only a few mutants available for these cells, and the culturing of these cells is much more difficult and slower than for bacteria. Nevertheless, great progress has been made in our understanding of DNA repair and mutagenesis in both yeast and human cells.

Mammalian cells have essentially the same repair systems that were first discovered in bacteria, however, repair is more complicated in mammalian cells because the structure of the genome is more complicated. In bacteria, the DNA is essentially naked, while the DNA in mammalian cells is surrounded by proteins. These proteins must first be removed before DNA repair can proceed. Therefore, more genes are involved in nucleotide excision repair in mammalian cells than in bacteria.

DNA repair mutants in yeast can be produced in a manner similar to the production of mutants in bacteria, i.e., treat the cells with a mutagen and then select for the phenotype (characteristic) that you want, i.e., sensitivity to radiation. The mutants in human cells have largely come from patients who have shown a sensitivity to sunlight (e.g., Xeroderma pigmentosum) or to ionizing radiation, detected during radiation therapy for cancer (e.g., Ataxia telangectasia).

Through the use of such mutant cells, the pathways of DNA repair in mammalian cells have been discovered and studied in the same general manner that was used in the bacterial studies.

3

Photosynthesis

INTRODUCTION

Photosynthesis is a metabolic pathway that converts light energy into chemical energy. Its initial substrates are carbon dioxide and water; the energy source is sunlight (electromagnetic radiation); and the end-products are oxygen and (energy-containing) carbohydrates, such as sucrose, glucose or starch. This process is one of the most important biochemical pathways, since nearly all life on Earth either directly or indirectly depends on it as a source of energy. It is a complex process occurring in plants, algae, as well as bacteria such as cyanobacteria. Photosynthetic organisms are also referred to as photoautotrophs.

Photosynthesis splits water to liberate O_2 and fixes CO_2 into sugar.

Photosynthesis uses light energy and carbon dioxide to make triose phosphates (G3P). The G3P is generally considered the first end-product of photosynthesis. It can be used as a source of metabolic energy, or combined and rearranged to form monosaccharide or disaccharide sugars, such as glucose or sucrose, respectively, which can be transported to other cells, stored as insoluble polysaccharides such as starch, or converted to structural carbohydrates, such as cellulose or glucans.

A commonly used slightly simplified equation for photosynthesis is:

$6\ CO_2(g) + 12\ H_2O(l)$ + photons ? $C_6H_{12}O_6(aq) + 6\ O_2(g) + 6\ H_2O(l)$

carbon dioxide + water + light energy ? glucose + oxygen + water

The equation is often presented in introductory chemistry texts in an even more simplified form as

$6\ CO_2(g) + 6\ H_2O(l)$ + photons ? $C_6H_{12}O_6(aq) + 6\ O_2(g)$

Photosynthesis occurs in two stages. In the first stage, light-dependent reactions or photosynthetic reactions (also called the Light Reactions) capture the energy of light and use it to make high-energy molecules. During the second stage, the light-independent reactions (also called the Calvin-Benson Cycle, and formerly known as the Dark Reactions) use the high-energy molecules to capture and chemically reduce carbon dioxide (CO_2) (also called carbon fixation) to make the precursors of carbohydrates.

In the light reactions, one molecule of the pigment chlorophyll absorbs one photon and loses one electron. This electron is passed to a modified form of chlorophyll called pheophytin, which passes the electron to a quinone molecule, allowing the start of a flow of electrons down an electron transport chain that leads to the ultimate reduction of NADP to NADPH. In addition, this creates a proton gradient across the chloroplast membrane; its dissipation is used by ATP Synthase for the concomitant synthesis of ATP. The chlorophyll molecule regains the lost electron from a water molecule through a process called photolysis, which releases a dioxygen (O_2) molecule.

In the Light-independent or dark reactions the enzyme RuBisCO captures CO_2 from the atmosphere and in a process that requires the newly formed NADPH, called the Calvin-Benson Cycle, releases three-carbon sugars, which are later combined to form sucrose and starch.

Photosynthesis may simply be defined as the conversion of light energy into chemical energy by living organisms. It is affected by its surroundings, and the rate of photosynthesis is affected by the concentration of carbon dioxide in the air, the light intensity, and the temperature.

Photosynthesis uses only 1% of the entire electromagnetic spectrum, and 2% of the visible spectrum. It has been estimated that the productivity of photosythesis is 115 petagrams (Pg, equals 1015 grams or 109 metric tons).

PHOTOAUTOTROPHS

Most plants are photoautotrophs, which means that they are able to synthesize food directly from inorganic compounds using light energy - for example from the sun, instead of eating other organisms or relying on nutrients derived from them. This is distinct from chemoautotrophs that do not depend on light energy, but use energy from inorganic compounds.

$$6\ CO_2 + 12\ H_2O\ ?\ C_6H_{12}O_6 + 6\ O_2 + 6\ H_2O$$

The energy for photosynthesis ultimately comes from absorbed photons and involves a reducing agent, which is water in the case of plants, releasing oxygen as product. The light energy is converted to chemical energy (known as light-dependent reactions), in the form of ATP and NADPH, which are used for synthetic reactions in photoautotrophs. The overall equation for the light-dependent reactions under the conditions of non-cyclic electron flow in green plants is

$$2\ H_2O + 2\ NADP^+ + 2\ ADP + 2\ Pi + light\ ?\ 2\ NADPH + 2\ H^+ + 2\ ATP + O_2$$

Most notably, plants use the chemical energy to fix carbon dioxide into carbohydrates and other organic compounds through light-independent reactions. The overall equation for carbon fixation (sometimes referred to as carbon reduction) in green plants is:

$$3\ CO_2 + 9\ ATP + 6\ NADPH + 6\ H+\ ?\ C_3H_6O_3\text{-phosphate} + 9\ ADP + 8\ Pi + 6\ NADP+ + 3\ H_2O$$

To be more specific, carbon fixation produces an intermediate product, which is then converted to the final carbohydrate products. The carbon skeletons produced by photosynthesis are then variously used to form other organic compounds, such as the building material cellulose, as precursors for lipid and amino acid biosynthesis, or as a fuel in cellular respiration. The latter occurs not only in plants but also in animals when the energy from plants gets passed through a food chain. Organisms dependent on photosynthetic and chemosynthetic organisms are called heterotrophs. In general outline, cellular respiration is the opposite of photosynthesis: Glucose and other compounds are oxidized to produce carbon dioxide, water, and chemical energy. However, the two processes take place through a different sequence of chemical reactions and in different cellular compartments.

Plants absorb light primarily using the pigment chlorophyll, which is the reason that most plants have a green colour. The function of chlorophyll is often supported by other accessory pigments such as carotenes and xanthophylls. Both chlorophyll and accessory pigments are contained in organelles (compartments within the cell) called chloroplasts. Although all cells in the green parts of a plant have chloroplasts, most of the energy is captured in the leaves. The cells in the interior tissues of a leaf, called the mesophyll, can contain between 450,000 and 800,000 chloroplasts for every square millimeter of leaf. The surface of the leaf is uniformly coated with a water-resistant waxy cuticle that protects the leaf from excessive evaporation of water and decreases the absorption of ultraviolet or blue light to reduce heating. The transparent epidermis layer allows light to pass through to the palisade mesophyll cells where most of the photosynthesis takes place.

Plants convert light into chemical energy with a maximum photosynthetic efficiency of approximately 6% By comparison solar panels convert light into electric energy at a photosynthetic efficiency of approximately 10-20%. Actual plant's photosynthetic efficiency varies with the frequency of the light being converted, light intensity, temperature and proportion of CO_2 in atmosphere.

ALGAE AND BACTERIA

Algae come in multiple forms from multicellular organisms like kelp, to microscopic, single-cell organisms. Although they are not as complex as land plants, the biochemical process of photosynthesis is the same. Very much like plants, algae have chloroplasts and chlorophyll, but various accessory pigments are present in some algae such as phycocyanin, carotenes, and xanthophylls in green algae and phycoerythrin in red algae (rhodophytes), resulting in a wide variety of colors. Brown algae and diatoms contain fucoxanthol as their primary pigment. All algae produce oxygen, and many are autotrophic. However, some are heterotrophic, relying on materials produced by other organisms. For example, in coral reefs, there is a mutualistic relationship between zooxanthellae and the coral polyps.

Photosynthetic bacteria do not have chloroplasts (or any membrane-bound organelles). Instead, photosynthesis takes place directly within the cell. Cyanobacteria contain thylakoid membranes very similar to those in chloroplasts and are the only prokaryotes that perform oxygen-generating photosynthesis. In fact, chloroplasts are now considered to have evolved from an endosymbiotic bacterium, which was also an ancestor of cyanobacterium.

The other photosynthetic bacteria have a variety of different pigments, called bacteriochlorophylls, and use electron donors different from water and thus do not produce oxygen. Some bacteria, such as Chromatium, oxidize hydrogen sulfide instead of water for photosynthesis, producing sulfur as waste. Other photosynthetic bacteria oxidize ferrous iron to ferric iron, others nitrite to nitrate and still others use arsenites, producing arsenates.

All photosynthesizing organisms must be in the photic (light-receiving) zone, except for those near hydrothermal vents which give faint light.

Some sea slugs have acquired the genetic machinery necessary for photosynthesis by horizontal gene transfer from their algal diet.

PLANT CELLS WITH VISIBLE CHLOROPLASTS

Early photosynthetic systems, such as those from green and purple sulfur and green and purple non-sulfur bacteria, are thought to have been anoxygenic, using various molecules as electron donors. Green and purple sulfur bacteria are thought to have used hydrogen and sulfur as an electron donor. Green nonsulfur bacteria used various amino and other organic acids. Purple nonsulfur bacteria used a variety of non-specific organic molecules. The use of these molecules is consistent with the geological evidence that the atmosphere was highly reduced at that time.

Fossils of what are thought to be filamentous photosynthetic organisms have been dated at 3.4 billion years old.

The main source of oxygen in the atmosphere is oxygenic photosynthesis, and its first appearance is sometimes referred to as the oxygen catastrophe. Geological evidence suggests that oxygenic photosynthesis, such as that in cyanobacteria, became important during the Paleoproterozoic era around 2 billion years ago. Modern photosynthesis in plants and most photosynthetic prokaryotes is oxygenic. Oxygenic photosynthesis uses water as an electron donor which is oxidized to molecular dioxygen (O_2) in the photosynthetic reaction center.

ORIGIN OF CHLOROPLASTS

In plants photosynthesis occurs in organelles called chloroplasts. Chloroplasts have many similarities with photosynthetic bacteria including a circular chromosome, prokaryotic-type ribosomes, and similar proteins in the photosynthetic reaction center.

The endosymbiotic theory suggests that photosynthetic bacteria were acquired (by endocytosis) by early eukaryotic cells to form the first plant cells. Therefore, chloroplasts may be photosynthetic bacteria that adapted to life inside plant cells. Like mitochondria, chloroplasts still possess their own DNA, separate from the nuclear DNA of their plant host cells and the genes in this chloroplast DNA resemble those in cyanobacteria.

Marine molluscs Elysia viridis and Elysia chlorotica likewise maintain a symbiotic relationship with chloroplasts that they capture from the algae that they ingest. This allows the molluscs to survive solely by photosynthesis for several months at a time.

CYANOBACTERIA AND THE EVOLUTION OF PHOTO-SYNTHESIS

The biochemical capacity to use water as the source for electrons in photosynthesis evolved once, in a common ancestor of extant cyanobacteria. The geological record indicates that this transforming event took place early in Earth's history, at least 2450-2320 million years ago (Ma), and possibly much earlier. Available evidence from geobiological studies of Archean (>2500 Ma) sedimentary rocks indicates that life existed 3500 Ma, but the question of when oxygenic photosynthesis evolved is still unanswered. A clear paleontological window on cyanobacterial evolution opened about 2000 Ma, revealing an already-diverse biota of blue-greens. Cyanobacteria remained principal primary producers throughout the Proterozoic Eon (2500-543 Ma), in part because the redox structure of the oceans favoured photo-autotrophs capable of nitrogen fixation. Green algae joined blue-greens as major primary producers on continental shelves near the end of the Proterozoic, but only with the Mesozoic (251-65 Ma) radiations of dinoflagellates, coccolithophorids, and diatoms did primary production in marine shelf waters take modern form. Cyanobacteria remain critical to marine ecosystems as primary producers in oceanic gyres, as agents of biological nitrogen fixation, and, in modified form, as the plastids of marine algae.

TEMPORAL ORDER

The overall process of photosynthesis takes place in four stages. The first, energy transfer in antenna chlorophyll takes place in the femtosecond [1 femtosecond (fs) = 10,-15 s] to picosecond [1 picosecond (ps) = 10-12 s] time scale. The next phase, the transfer of electrons in photochemical reactions, takes place in the picosecond to nanosecond time scale [1 nanosecond (ns) = 10-9 s]. The third phase, the electron transport chain and ATP synthesis, takes place on the microsecond [1 microsecond (μs) = 10-6 s] to millisecond [1 millisecond (ms) = 10-3 s) time scale. The final phase is carbon fixation and export of stable products and takes place in the millisecond to second time scale. The first three stages occur in the thylakoid membranes.

LIGHT TO CHEMICAL ENERGY

The light energy is converted to chemical energy using the light-dependent reactions. This chemical energy production is about 5-6% efficient, with the majority of the light that strikes a plant reflected and not absorbed. However, of the energy that is absorbed, approximately 30-50% is captured as chemical energy. The products of the light-dependent reactions are ATP from photophosphorylation and NADPH from photoreduction. Both are then utilized as an energy source for the light-independent reactions.

Not all wavelengths of light can support photosynthesis. The photosynthetic action spectrum depends on the type of accessory pigments present. For example, in green plants, the action spectrum resembles the absorption spectrum for chlorophylls and carotenoids with peaks for violet-blue and red light. In red algae, the action spectrum overlaps with the absorption spectrum of phycobilins for blue-green light, which allows these algae to grow in deeper waters that filter out the longer wavelengths used by green plants. The non-absorbed part of the light spectrum is what gives photosynthetic organisms their color (e.g., green plants, red algae, purple bacteria) and is the least effective for photosynthesis in the respective organisms.

In plants, light-dependent reactions occur in the thylakoid membranes of the chloroplasts and use light energy to synthesize ATP and NADPH. The light-dependent reaction has two forms; cyclic and non-cyclic reaction. In the non-cyclic reaction, the photons are captured in the light-harvesting antenna complexes of photosystem II by chlorophyll and other accessory pigments (see diagram at right). When a chlorophyll molecule at the core of the photosystem II reaction center obtains sufficient excitation energy from the adjacent antenna pigments, an electron is transferred to the primary electron-acceptor molecule, Pheophytin, through a process called Photoinduced charge separation. These electrons are shuttled through an electron transport chain, the so called Z-scheme, that initially functions to generate a chemiosmotic potential across the membrane. An ATP synthase enzyme uses the chemiosmotic potential to make

ATP during photophosphorylation, whereas NADPH is a product of the terminal redox reaction in the Z-scheme. The electron enters the Photosystem I molecule. The electron is excited due to the light absorbed by the photosystem. A second electron carrier accepts the electron, which again is passed down lowering energies of electron acceptors. The energy created by the electron acceptors is used to move hydrogen ions across the thylakoid membrane into the lumen. The electron is used to reduce the co-enzyme NADP, which has functions in the light-independent reaction. The cyclic reaction is similar to that of the non-cyclic, but differs in the form that it generates only ATP, and no reduced NADP (NADPH) is created. The cyclic reaction takes place only at photosystem I. Once the electron is displaced from the photosystem, the electron is passed down the electron acceptor molecules and returns back to photosystem I, from where it was emitted, hence the name cyclic reaction.

WATER PHOTOLYSIS

The NADPH is the main reducing agent in chloroplasts, providing a source of energetic electrons to other reactions. Its production leaves chlorophyll with a deficit of electrons (oxidized), which must be obtained from some other reducing agent. The excited electrons lost from chlorophyll in photosystem I are replaced from the electron transport chain by plastocyanin. However, since photosystem II includes the first steps of the Z-scheme, an external source of electrons is required to reduce its oxidized chlorophyll a molecules. The source of electrons in green-plant and cyanobacterial photosynthesis is water. Two water molecules are oxidized by four successive charge-separation reactions by photosystem II to yield a molecule of diatomic oxygen and four hydrogen ions; the electron yielded in each step is transferred to a redox-active tyrosine residue that then reduces the photoxidized paired-chlorophyll a species called P680 that serves as the primary (light-driven) electron donor in the photosystem II reaction center. The oxidation of water is catalyzed in photosystem II by a redox-active structure that contains four manganese ions and a calcium ion; this oxygen-evolving complex binds two water molecules and stores the four oxidizing equivalents that are required to drive the water-oxidizing reaction. Photosystem II is the only known biological enzyme that carries out this oxidation of water. The hydrogen ions contribute to the transmembrane chemiosmotic potential that leads to ATP synthesis. Oxygen is a waste product of light-independent reactions, but the majority of organisms on Earth use oxygen for cellular respiration, including photosynthetic organisms.

QUANTUM MECHANICAL EFFECTS

Through photosynthesis, sunlight energy is transferred to molecular reaction centers for conversion into chemical energy with nearly 100-per cent efficiency. The transfer of the solar energy takes place almost instantaneously, so little energy is wasted as heat. Of the total incident solar radiation only 43% can be used (only light in the range 400-700 nm), 80% of light makes it through the canopy, photosynthesis stores 28.6% of the energy, and plant respiration uses some energy which leaves 67% of the stored energy behind. This brings the actual efficiency of photosynthesis to about 6.6%

A study led by researchers with the U.S. Department of Energy's Lawrence Berkeley National Laboratory (Berkeley Lab) and the University of California at Berkeley suggests that long-lived wavelike electronic quantum coherence plays an important part in this instantaneous transfer of energy by allowing the photosynthetic system to simultaneously try each potential energy pathway and choose the most efficient option. Results of the study are presented in the April 12, 2007 issue of the journal Nature.

OXYGEN AND PHOTOSYNTHESIS

With respect to oxygen and photosynthesis, there are two important concepts.

Plant and cyanobacterial (blue-green algae) cells also use oxygen for cellular respiration, although they have a net output of oxygen since much more is produced during photosynthesis.

Oxygen is a product of the light-driven water-oxidation reaction catalyzed by photosystem II; it is not generated by the fixation of carbon dioxide. Consequently, the source of oxygen during photosynthesis is water, not carbon dioxide.

Bacterial Variation

The concept that oxygen production is not directly associated with the fixation of carbon dioxide was first proposed by Cornelis Van Niel in the 1930s, who studied photosynthetic bacteria. Aside from the cyanobacteria, bacteria only have one photosystem and use reducing agents other than water. They get electrons from a variety of different inorganic chemicals including sulfide or hydrogen, so for most of these bacteria oxygen is not produced.

Others, such as the halophiles (an Archaea), produced so-called purple membranes where the bacteriorhodopsin could harvest light and produce energy. The purple membranes was one of the first to be used to demonstrate the chemiosmotic theory: light hit the membranes and the pH of the solution that contained the purple membranes dropped as protons were pumping out of the membrane.

The fixation or reduction of carbon dioxide is a light-independent process in which carbon dioxide combines with a five-carbon sugar, ribulose 1,5-bisphosphate (RuBP), to yield two molecules of a three-carbon compound, glycerate 3-phosphate (GP), also known as 3-phosphoglycerate (PGA). GP, in the presence of ATP and NADPH from the light-dependent stages, is reduced to glyceraldehyde 3-phosphate (G3P). This product is also referred to as 3-phosphoglyceraldehyde (PGAL) or even as triose phosphate. Triose is a 3-carbon sugar (see carbohydrates). Most (5 out of 6 molecules) of the G3P produced is used to regenerate RuBP so the process can continue (see Calvin-Benson cycle). The 1 out of 6 molecules of the triose phosphates not "recycled" often condense to form hexose phosphates, which ultimately yield sucrose, starch and cellulose. The sugars produced during carbon metabolism yield carbon skeletons that can be used for other metabolic reactions like the production of amino acids and lipids.

In hot and dry conditions, plants will close their stomata to prevent loss of water. Under these conditions, CO_2 will decrease, and dioxygen gas, produced by the light reactions of photosynthesis, will increase in the leaves, causing an increase of photorespiration by the oxygenase activity of ribulose-1,5-bisphosphate carboxylase/oxygenase and decrease in carbon fixation. Some plants have evolved mechanisms to increase the CO_2 concentration in the leaves under these conditions.

C_4 plants chemically fix carbon dioxide in the cells of the mesophyll by adding it to the three-carbon molecule phosphoenolpyruvate (PEP), a reaction catalyzed by an enzyme called PEP carboxylase and which creates the four-carbon organic acid, oxaloacetic acid. Oxaloacetic acid or malate synthesized by this process is then translocated to specialized bundle sheath cells where the enzyme, rubisco, and other Calvin cyle enzymes are located, and where CO_2 released by decarboxylation of the four-carbon acids is then fixed by rubisco activity to the three-carbon sugar 3-phosphoglycerate. The physical separation of rubisco from the oxygen-generating light reactions reduces photorespiration and increases CO_2 fixation and thus photosynthetic capacity of the leaf. C_4 plants can produce more sugar than C_3 plants in conditions of high light and temperature. Many important crop plants are C_4 plants including maize, sorghum, sugarcane, and millet. Plants lacking PEP-carboxylase are called C_3 plants because the primary carboxylation reaction, catalyzed by Rubiso, produces the three-carbon sugar 3-phosphoglycerate directly in the Calvin-Benson Cycle.

CAM PHOTOSYNTHESIS

Xerophytes such as cacti and most succulents also use PEP carboxylase to capture carbon dioxide in a process called Crassulacean acid metabolism (CAM). In contrast to C_4 metabolism, which physically separates the CO_2 fixation to PEP from the Calvin cycle, CAM only temporally separates these two processes. CAM plants have a different leaf anatomy than C_4 plants, and fix the CO_2 at night, when their stomata are open.

CAM plants store the CO_2 mostly in the form of malic acid via carboxylation of phosphoenolpyruvate to oxaloacetate, which is then reduced to malate. Decarboxylation of malate during the day releases CO_2 inside the leaves thus allowing carbon fixation to 3-phosphoglycerate by rubisco.

Although some of the steps in photosynthesis are still not completely understood, the overall photosynthetic equation has been known since the 1800s.

Jan van Helmont began the research of the process in the mid-1600s when he carefully measured the mass of the soil used by a plant and the mass of the plant as it grew. After noticing that the soil mass changed very little, he hypothesized that the mass of the growing plant must come from the water, the only substance he added to the potted plant. His hypothesis was partially accurate - much of the gained mass also comes from carbon dioxide as well as water. However, this was a signaling point to the idea that the bulk of a plant's biomass comes from the inputs of photosynthesis, not the soil itself.

Joseph Priestley, a chemist and minister, discovered that when he isolated a volume of air under an inverted jar, and burned a candle in it, the candle would burn out very quickly, much before it ran out of wax. He further discovered that a mouse could similarly "injure" air. He then showed that the air that had been "injured" by the candle and the mouse could be restored by a plant.

In 1778, Jan Ingenhousz, court physician to the Austrian Empress, repeated Priestley's experiments. He discovered that it was the influence of sunlight on the plant that could cause it to rescue a mouse in a matter of hours.

In 1796, Jean Senebier, a Swiss pastor, botanist, and naturalist, demonstrated that green plants consume carbon dioxide and release oxygen under the influence of light. Soon afterwards, Nicolas-Théodore de Saussure showed that the increase in mass of the plant as it grows could not be due only to uptake of CO_2, but also to the incorporation of water. Thus the basic reaction by which photosynthesis is used to produce food (such as glucose) was outlined.

Cornelis Van Niel made key discoveries explaining the chemistry of photosynthesis. By studying purple sulfur bacteria and green bacteria he was the first scientist to demonstrate that photosynthesis is a light-dependent redox reaction, in which hydrogen reduces carbon dioxide.

Robert Emerson discovered two light reactions by testing plant productivity using different wavelengths of light. With the red alone, the light reactions were suppressed. When blue and red were combined, the output was much more substantial. Thus, there were two photosystems, one aborbing up to 600 nm wavelengths, the other up to 700. The former is known as PSII, the latter is PSI. PSI contains only chlorophyll a, PSII contains primarily chlorophyll a with most of the available chlorophyll b, among other pigments.

Further experiments to prove that the oxygen developed during the photosynthesis of green plants came from water, were performed by Robert Hill in 1937 and 1939. He showed that isolated chloroplasts give off oxygen in the presence of unnatural reducing agents like iron oxalate, ferricyanide or benzoquinone after exposure to light. The Hill reaction is as follows:

$$2\,H_2O + 2\,A + \text{(light, chloroplasts)}\ ?\ 2\,AH_2 + O_2$$

where A is the electron acceptor. Therefore, in light the electron acceptor is reduced and oxygen is evolved. Cyt b6, now known as a plastoquinone, is one electron acceptor.

Samuel Ruben and Martin Kamen used radioactive isotopes to determine that the oxygen liberated in photosynthesis came from the water.

Melvin Calvin and Andrew Benson, along with James Bassham, elucidated the path of carbon assimilation (the photosynthetic carbon reduction cycle) in plants. The carbon reduction cycle is known as the Calvin cycle, which inappropriately ignores the contribution of Bassham and Benson. Many scientists refer to the cycle as the Calvin-Benson Cycle, Benson-Calvin, and some even call it the Calvin-Benson-Bassham (or CBB) Cycle.

A Nobel Prize winning scientist, Rudolph A. Marcus, was able to discover the function and significance of the electron transport chain.

There are three main factors affecting photosynthesis and several corollary factors. The three main are:

- Light irradiance and wavelength
- Carbon dioxide concentration
- Temperature.

Light Intensity (irradiance), Wavelength and Temperature

In the early 1900s Frederick Frost Blackman along with Gabrielle Matthaei investigated the effects of light intensity (irradiance) and temperature on the rate of carbon assimilation.

At constant temperature, the rate of carbon assimilation varies with irradiance, initially increasing as the irradiance increases. However at higher irradiance this relationship no longer holds and the rate of carbon assimilation reaches a plateau.

At constant irradiance, the rate of carbon assimilation increases as the temperature is increased over a limited range. This effect is only seen at high irradiance levels. At low irradiance, increasing the temperature has little influence on the rate of carbon assimilation.

These two experiments illustrate vital points: firstly, from research it is known that photochemical reactions are not generally affected by temperature. However, these experiments clearly show that temperature affects the rate of carbon assimilation, so there must be two sets of reactions in the full process of carbon assimilation. These are of course the light-dependent 'photochemical' stage and the light-independent, temperature-dependent stage. Second, Blackman's experiments illustrate the concept of limiting factors. Another limiting factor is the wavelength of light. Cyanobacteria, which reside several meters underwater, cannot receive the correct wavelengths required to cause photoinduced charge separation in conventional photosynthetic pigments. To combat this problem, a series of proteins with different pigments surround the reaction center. This unit is called a phycobilisome.

Carbon Dioxide Levels and Photorespiration

As carbon dioxide concentrations rise, the rate at which sugars are made by the light-independent reactions increases until limited by other factors. RuBisCO, the enzyme that captures carbon dioxide in the light-independent reactions, has a binding affinity for both carbon dioxide and oxygen. When the concentration of carbon dioxide is high, RuBisCO will fix carbon dioxide. However, if the oxygen concentration is high, RuBisCO will bind oxygen instead of carbon dioxide. This process, called photorespiration, uses energy, but does not make sugar.

RuBisCO oxygenase activity is disadvantageous to plants for several reasons:

One product of oxygenase activity is phosphoglycolate (2 carbon) instead of 3-phosphoglycerate (3 carbon). Phosphoglycolate cannot be metabolized by the Calvin-Benson cycle and represents carbon lost from the cycle. A high oxygenase activity, therefore, drains the sugars that are required to recycle ribulose 5-bisphosphate and for the continuation of the Calvin-Benson cycle.

Phosphoglycolate is quickly metabolized to glycolate that is toxic to a plant at a high concentration; it inhibits photosynthesis.

Salvaging glycolate is an energetically expensive process that uses the glycolate pathway and only 75% of the carbon is returned to the Calvin-Benson cycle as 3-phosphoglycerate.

A highly-simplified summary is:

2 glycolate + ATP ? 3-phophoglycerate + carbon dioxide + ADP + NH3

The salvaging pathway for the products of RuBisCO oxygenase activity is more commonly known as photorespiration, since it is characterized by light-dependent oxygen consumption and the release of carbon dioxide.

4

Calvin Cycle

INTRODUCTION

The Calvin cycle (or Calvin-Benson-Bassham cycle or carbon fixation) is a series of biochemical reactions that takes place in the stroma of chloroplasts in photosynthetic organisms. It was discovered by Melvin Calvin, James Bassham and Andrew Benson at the University of California, Berkeley . It is one of the light-independent reactions or dark reactions.

During photosynthesis, light energy is used to generate chemical free energy, stored in glucose. The light-independent Calvin cycle, also (misleadingly) known as the "dark reaction" or "dark stage", uses the energy from short-lived electronically-excited carriers to convert carbon dioxide and water into organic compounds that can be used by the organism (and by animals which feed on it). This set of reactions is also called *carbon fixation.* The key enzyme of the cycle is called RuBisCO. In the following equations, the chemical species (phosphates and carboxylic acids) exist in equilibria among their various ionized states as governed by the pH.

The enzymes in the Calvin cycle are functionally equivalent to many enzymes used in other metabolic pathways such as gluconeogenesis and the pentose phosphate pathway, but they are to be found in the chloroplast stroma instead of the cell cytoplasm, separating the reactions. They are activated in the light (which is why the name "dark reaction" is misleading), and also by products of the light-dependent reaction. These regulatory functions prevent the Calvin cycle from operating in reverse to respiration, which would create a continuous cycle of carbon dioxide being reduced to carbohydrates, and carbohydrates being respired to carbon dioxide. Energy (in the form of ATP) would be wasted in carrying out these reactions that have no net productivity.

The sum of reactions in the Calvin cycle is the following:

$3\ CO_2 + 6\ NADPH + 5\ H_2O + 9\ ATP\ ?\ C_3H_5O_3\text{-}PO_3^{2-} + 2\ H^+ + 6\ NADP^+ + 9\ ADP + 8\ P_i$

OR

$3\ CO_2 + 6\ C_{21}H_{29}N_7O_{17}P_3 + 5\ H_2O + 9\ C_{10}H_{16}N_5O_{13}P_3\ ?\ C_3H_5O_3 - PO_3^{2-} + 2\ H^+ + 6\ NADP^+ + 9\ C_{10}H_{15}N_5O_{10}P_2 + 8\ P_i$

It should be noted that hexose (six carbon) sugars are not a product of the Calvin cycle. Although many texts list a product of photosynthesis as $C_6H_{12}O_6$, this is mainly a convenience to counter the equation of respiration, where six-carbon sugars are oxidized in mitochondria. The carbohydrate products of the Calvin Cycle are three-carbon sugar phosphate molecules, or "triose phosphates," specifically, glyceraldehyde-3-phosphate (G3P).

STEPS OF THE CALVIN CYCLE

The enzyme RuBisCO catalyses the carboxylation of Ribulose-1,5-bisphosphate, a 5 carbon compound, by carbon dioxide (a total of 6 carbons) in a two-step reaction. The initial product of the reaction is a six-carbon intermediate so unstable that it immediately splits in half, forming two molecules of glycerate 3-phosphate, a 3-carbon compound. (also: 3-phosphoglycerate, 3-phosphoglyceric acid, 3PGA)

The enzyme phosphoglycerate kinase catalyses the phosphorylation of 3PGA by ATP (which was produced in the light-dependent stage). 1,3-bisphosphoglycerate (glycerate-1,3-bisphosphate) and ADP are the products. (However, note that two PGAs are produced for every CO_2 that enters the cycle, so this step utilizes 2ATP per CO_2 fixed.

The enzyme G3P dehydrogenase catalyses the reduction of 1,3BPGA by NADPH (which is another product of the light-dependent stage). Glyceraldehyde 3-phosphate (also G3P, GP) is produced, and the NADPH itself was oxidized and becomes $NADP^+$. Again, two NADPH are utilized per CO_2 fixed.

Triose phosphate isomerase converts some G3P reversibly into dihydroxyacetone phosphate (DHAP), also a 3-carbon molecule.

Aldolase and fructose-1,6-bisphosphatase convert a G3P and a DHAP into fructose-6-phosphate (6C). A phosphate ion is lost into solution. Then fixation of another CO_2 generates two more G3P.

F6P has two carbons removed by transketolase, giving erythrose-4-phosphate. The two carbons on transketolase are added to a G3P, giving the ketose xylulose-5-phosphate (Xu5P).

E4P and a DHAP (formed from one of the G3P from the second CO_2 fixation) are converted into sedoheptulose-1,7-bisphosphate (7C) by aldolase enzyme.

Sedoheptulose-1,7-bisphosphatase (one of only three enzymes of the Calvin cycle which are unique to plants) cleaves sedoheptulose-1,7-bisphosphate into sedoheptulose-7-phosphate, releasing an inorganic phosphate ion into solution.

Fixation of a third CO_2 generates two more G3P. The ketose S7P has two carbons removed by transketolase, giving ribose-5-phosphate (R5P), and the two carbons remaining on transketolase are transferred to one of the G3P, giving another Xu5P. This leaves one G3P as the product of fixation of 3 CO_2, with generation of three pentoses which can be converted to Ru5P.

R5P is converted into ribulose-5-phosphate (Ru5P, RuP) by phosphopentose isomerase. Xu5P is converted into RuP by phosphopentose epimerase.

Finally, phosphoribulokinase (another plant unique enzyme of the pathway) phosphorylates RuP into RuBP, ribulose-1,5-bisphosphate, completing the Calvin *cycle*. This requires the input of one ATP.

Thus, of 6 G3P produced, three RuBP (5C) are made totalling 15 carbons, with only one available for subsequent conversion to hexose. This required 9 ATPs and 6 NADPH per 3 CO_2.

RuBisCO also reacts competitively with O_2 instead of CO_2 in *photorespiration*. The rate of photorespiration is higher at high temperatures. "photorespiration" turns RuBP into 3PGA and 2-phosphoglycolate, a 2-carbon molecule which can be converted via glycolate and glyoxalate to glycine. Via the glycine cleavage system and tetrahydrofolate, two glycines are converted into serine +CO_2. Serine can be converted back to 3-phosphoglycerate. Thus, only 3 of 4 carbons from two phosphoglycolates can be converted back to 3PGA. Obviously photorespiration has very negative consequences for the plant, because rather than fixing CO_2, this process leads to loss of CO_2. C4 carbon fixation evolved to circumvent photorespiration, but can only occur in certain plants living in very warm or tropical climates.

PRODUCTS OF THE CALVIN CYCLE

The immediate product of the Calvin cycle is glyceraldehyde-3-phosphate (G3P) and water. Two G3P molecules (or one F6P molecule) that have exited the cycle are used to make larger carbohydrates. In simplified versions of the Calvin cycle they may be converted to F6P or F5P after exit, but this conversion is also part of the cycle.

Hexose isomerase converts about half of the F6P molecules in to glucose-6-phosphate. These are dephosphorylated and the glucose can be used to form starch, which is stored in, for example, potatoes, or cellulose used to build up cell walls. Glucose, with fructose, forms sucrose, a non-reducing sugar which is a stable storage sugar, unlike glucose.

PHOTORESPIRATION

Photorespiration (or "photo-respiration") is the alternate pathway for production of glyceraldehyde 3-phosphate (G3P) by RuBisCO, the main enzyme of the light-independent reactions of photosynthesis (also known as the Calvin cycle or the Primary Carbon Reduction (PCR) cycle). Although RuBisCO favors carbon dioxide to oxygen,(approximately 3 carboxylations per oxygenation), oxygenation of RuBisCO occurs frequently, producing a glycolate and a glycerate. This usually occurs when oxygen levels are high; for example, when the stomata (tiny pores on the leaf) are closed to prevent water loss on dry days. It involves three cellular organelles: chloroplasts, peroxisomes, and mitochondria. Photorespiration produces no ATP.

Biochemistry of Photorespiration

The oxidative photosynthetic carbon cycle reaction is catalyzed by RuBP oxygenase activity:

RuBP + O_2 ? Phosphoglycolate + 3-phosphoglycerate

The phosphoglycolate is salvaged by a series of reactions in the peroxisome, mitochondria, and again in the peroxisome where it is converted into serine and later glycerate. Glycerate reenters the chloroplast and subsequently the Calvin cycle by the same transporter that exports glycolate. A cost of 1 ATP is associated with conversion to 3-phosphoglycerate (PGA) (Phosphorylation), within the chloroplast, which is then free to reenter the PCR cycle. One carbon dioxide molecule is produced for every 2 molecules of O_2 that are taken up by RuBisCO.

Photorespiration is a wasteful process because G3P is created at a reduced rate and higher metabolic cost (2ATP and one NAD(P)H) compared with RuBP carboxylase activity. G3P produced in the chloroplast is used to create "nearly all" of the food and structures in the plant. While photorespiratory carbon cycling results in G3P eventually, it also produces waste ammonia that must be detoxified at a substantial cost to the cell in ATP and reducing equivalents.

ROLE OF PHOTORESPIRATION

Photorespiration is said to be an evolutionary relic. Photorespiration lowers the efficiency of photosynthesis by removing carbon molecules from the Calvin Cycle. The early atmosphere in which primitive plants originated contained very little oxygen, so it is hypothesized that the early evolution of RuBisCO was not influenced by its lack of discrimination between O_2 and carbon dioxide.

Another theory postulates that it may function as a "safety valve", preventing excess NADPH and ATP from reacting with oxygen and producing free radicals, as these can damage the metabolic functions of the cell by subsequent reactions with lipids or metabolites of alternate pathways.

MINIMIZATION OF PHOTORESPIRATION (C4 AND CAM PLANTS)

Maize uses the C4 pathway, minimizing photorespiration.

Since photorespiration requires additional energy from the light reactions of photosynthesis, some plants have mechanisms to reduce uptake of molecular oxygen by RuBisCO. They increase the concentration of CO_2 in the leaves so that Rubisco is less likely to produce glycolate through reaction with O_2.

C4 plants capture carbon dioxide in cells of their mesophyll (using an enzyme called PEP carboxylase), and oxaloacetate is formed. This oxaloacetate is then converted to malate and is release into the bundle sheath cells (site of carbon dioxide fixation by RuBisCO) where oxygen concentration is low to avoid photorespiration. Here Carbon dioxide is removed from the malate and combined with RuBP in the usual way. The Calvin cycle then proceeds as normal.

The enzyme PEP carboxylase (which catalyzes the combination of carbon dioxide with a compound called Phosphoenolpyruvate or PEP) is also found in other plants such as cacti and succulents who use a mechanism called Crassulacean acid metabolism or CAM in which PEP carboxylase sequesters carbon at night and releases it to the photosynthesizing cells during the day. This provides a mechanism for reducing high rates of water loss (transpiration) by stomata during the day.

This ability to avoid photorespiration makes these plants more hardy than other plants in dry conditions where stomata are closed and oxygen concentrations rise. C4 plants include sugar cane, corn (maize), and sorghum.

5

Cellular Respiration

Cellular respiration is the set of the metabolic reactions and processes that take place in organisms' cells to convert biochemical energy from nutrients into adenosine triphosphate (ATP), and then release waste products. The reactions involved in respiration are catabolic reactions that involve the oxidation of one molecule and the reduction of another.

Nutrients commonly used by animal and plant cells in respiration include glucose, amino acids and fatty acids, and a common oxidizing agent (electron acceptor) is molecular oxygen (O_2). Bacteria and archaea can also be lithotrophs and these organisms may respire using a broad range of inorganic molecules as electron donors and acceptors, such as sulfur, metal ions, methane or hydrogen. Organisms that use oxygen as a final electron acceptor in respiration are described as aerobic, while those that do not are referred to as anaerobic.

The energy released in respiration is used to synthesize ATP to store this energy. The energy stored in ATP can then be used to drive processes requiring energy, including biosynthesis, locomotion or transportation of molecules across cell membranes. Because of its ubiquity in nature, ATP is also known as the "universal energy currency".

AEROBIC RESPIRATION

Aerobic respiration requires oxygen in order to generate energy (ATP). It is the preferred method of pyruvate breakdown from glycolysis and requires that pyruvate enter the mitochondrion in order to be fully oxidized by the Krebs cycle. The product of this process is energy in the form of ATP (Adenosine Triphosphate), by substrate-level phosphorylation, NADH and FADH2.

Simplified Reaction

$C_6H_{12}O_{6\,(aq)} + 6O_{2\,(g)}$? $6CO_{2\,(g)} + 6H_2O_{\,(l)}$?G = -2880 kJ per mole of $C_6H_{12}O_6$

The reducing potential of NADH and FADH2 is converted to more ATP through an electron transport chain with oxygen as the "terminal electron acceptor". Most of the ATP produced by aerobic cellular respiration is made by oxidative phosphorylation. This works by the energy released in the consumption of pyruvate being used to create a chemiosmotic potential by pumping protons across a membrane. This potential is then used to drive ATP synthase and produce ATP from ADP. Biology textbooks often state that between 36-38 ATP molecules can be made per oxidised glucose molecule during cellular respiration (2 from glycolysis, 2 from the Krebs cycle, and about 32-34 from the electron transport system). Generally, 38 ATP molecules are formed from aerobic respiration. However, this maximum yield is never quite reached due to losses (leaky membranes) as well as the cost of moving pyruvate and ADP into the mitochondrial matrix.

Aerobic metabolism is 19 times more efficient than anaerobic metabolism (which yields 2 mol ATP per 1 mol glucose). They share the initial pathway of glycolysis but aerobic metabolism continues with the Krebs cycle and oxidative phosphorylation. The post glycolytic reactions take place in the mitochondria in eukaryotic cells, and in the cytoplasm in prokaryotic cells.

Glycolysis

Glycolysis is a metabolic pathway that is found in the cytoplasm of cells in all living organisms and is anaerobic, or doesn't require oxygen. The process converts one molecule of glucose into two molecules of pyruvate, and makes energy in the form of two net molecules of ATP. Four molecules of ATP per glucose are actually produced; however, two are consumed for the preparatory phase. The initial phosphorylation of glucose is required to destabilize the molecule for cleavage into two triose sugars. During the pay-off phase of glycolysis, four phosphate groups are transferred to ADP by substrate-level phosphorylation to make four ATP, and two NADH are produced when the triose sugars are oxidized. The overall reaction can be expressed this way:

$$\text{Glucose} + 2\ NAD^+ + 2\ P_i + 2\ ADP\ ?\ 2\ \text{pyruvate} + 2\ NADH + 2\ ATP + 2H^+ + 2\ H_2O$$

Oxidative Decarboxylation of Pyruvate

The pyruvate is oxidized to acetyl-CoA and CO_2 by the Pyruvate dehydrogenase complex, a cluster of enzymes—multiple copies of each of three enzymes—located in the mitochondria of eukaryotic cells and in the cytosol of prokaryotes. In the process one molecule of NADH is formed per pyruvate oxidized, and 3 moles of ATP are formed for each mole of pyruvate. This step is also known as the *link reaction*, as it links glycolysis and the Krebs cycle.

This is also called the *Krebs cycle* or the *tricarboxylic acid cycle*. When oxygen is present, acetyl-CoA is produced from the pyruvate molecules created from glycolysis. Once Acetyl CoA is formed, two processes can occur, aerobic or anaerobic respiration. When oxygen is present, the mitochondria will undergo aerobic respiration which leads to the Krebs cycle. However, if oxygen is not present, fermentation of the pyruvate molecule will occur. In the presence of oxygen, when acetyl-CoA is produced, the molecule then enters the citric acid cycle (Krebs cycle) inside the mitochondrial matrix, and gets oxidized to CO_2 while at the same time reducing NAD to NADH. NADH can be used by the electron transport chain to create further ATP as part of oxidative phosphorylation. To fully oxidize the equivalent of one glucose molecule, two acetyl-CoA must be metabolized by the Krebs cycle. Two waste products, H_2O and CO_2, are created during this cycle.

The citric acid cycle is an 8-step process involving 8 different enzymes. Throughout the entire cycle, Acetyl CoA changes into Citrate, Isocitrate, a-ketoglutarate, succinyl-CoA, succinate, fumarate, malate, and finally, oxaloacetate. The net energy gain from one cycle is 3 NADH, 1 FADH, and 1 GTP. Thus, the total amount of energy yield from one whole glucose molecule (2 pyruvate molecules) is 6 NADH, 2 FADH, and 2 ATP.

Oxidative Phosphorylation

In eukaryotes, oxidative phosphorylation occurs in the mitochondrial cristae. It comprises the electron transport chain that establishes a proton gradient (chemiosmotic potential) across the inner membrane by oxidizing the NADH produced from the Krebs cycle. ATP is synthesised by the ATP synthase enzyme when the chemiosmotic gradient is used to drive the phosphorylation of ADP. The electrons are finally transferred to exogenous oxygen, and with the addition of two protons, water is formed.

Although there is a theoretical yield of 36-38 ATP molecules per glucose during cellular respiration, such conditions are generally not realized due to losses such as the cost of moving pyruvate (from glycolysis), phosphate, and ADP (substrates for ATP synthesis) into the mitochondria. All are actively transported using carriers that utilise the stored energy in the proton electrochemical gradient.

- Pyruvate is taken up by a specific, low km transporter to bring it into the mitochondrial matrix for oxidation by the pyruvate dehydrogenase complex.
- The phosphate translocase is a symporter and the driving force for moving phosphate ions into the mitochondria is the proton motive force.
- The adenine nucleotide carrier is an antiporter and exchanges ADP and ATP across the inner membrane. The driving force is due to the ATP (-4) having a more negative charge than the ADP (-3) and thus it dissipates some of the electrical component of the proton electrochemical gradient.

The outcome of these transport processes using the proton electrochemical gradient is that more than 3 H^+ are needed to make 1 ATP. Obviously this reduces the theoretical efficiency of the whole process and the likely maximum is closer to 28-30 ATP molecules. In practice the efficiency may be even lower due to the inner membrane of the mitochondria being slightly leaky to protons. Other factors may also dissipate the proton gradient creating an apparently leaky mitochondria. An uncoupling protein known as thermogenin is expressed in some cell types and is a channel that can transport protons. When this protein is active in the inner membrane it short circuits the coupling between the electron transport chain and ATP synthesis. The potential energy from the proton gradient is not used to make ATP but generates heat. This is particularly important in a baby's brown fat, for thermogenesis, and hibernating animals.

Anaerobic Respiration

Without oxygen, pyruvate is not metabolized by cellular respiration but undergoes a process of fermentation. The pyruvate is not transported into the mitochondrion, but remains in the cytoplasm, where it is converted to waste products that may be removed from the cell. This serves the purpose of oxidizing the hydrogen carriers so that they can perform glycolysis again and removing the excess pyruvate. This waste product varies depending on the organism. In skeletal muscles, the waste product is lactic acid. This type of fermentation is called lactic acid fermentation. In yeast, the waste products are ethanol and carbon dioxide. This type of fermentation is known as alcoholic or ethanol fermentation. The ATP generated in this process is made by *substrate phosphorylation*, which is phosphorylation that does not involve oxygen.

Anaerobic respiration is less efficient at using the energy from glucose since 2 ATP are produced during anaerobic respiration per glucose, compared to the 36 ATP per glucose produced by aerobic respiration. This is because the waste products of anaerobic respiration still contain plenty of energy. Ethanol, for example, can be used in gasoline (petrol) solutions. Glycolytic ATP, however, is created more quickly. For prokaryotes to continue a rapid growth rate when they are shifted from an aerobic environment to an anaerobic environment, they must increase the rate of the glycolytic reactions. Thus, during short bursts of strenuous activity, muscle cells use anaerobic respiration to supplement the ATP production from the slower aerobic respiration, so anaerobic respiration may be used by a cell even before the oxygen levels are depleted, as is the case in sports that do not require athletes to pace themselves, such as sprinting.

CHEMOSYNTHESIS

Chemosynthesis is the biological conversion of one or more carbon molecules (usually carbon dioxide or methane) and nutrients into organic matter using the oxidation of inorganic molecules (e.g. hydrogen gas, hydrogen sulfide) or methane as a source of energy, rather than sunlight, as in photosynthesis. Chemoautotrophs, organisms that obtain carbon through chemosynthesis are phlogenetically diverse, but groups that include conspicuous or biogeochemically-important taxa include the sulfur-oxidizing gamma and epsilon proteobacteria, the Aquificaeles, the Methanogenic archaea and the neutrophilic iron-oxidizing bacteria.

Many microorganisms in dark regions of the oceans use chemosynthesis to produce biomass from 1-carbon molecules. Two categories can be distinguished. In the rare sites at which hydrogen molecules (H_2) are available, the energy available from the reaction between CO_2 and H_2 (leading to production of methane, CH_4) can be large enough to drive the production of biomass. Alternatively, in most oceanic environments, energy for chemosynthesis derives from reactions between O_2 and substances such as hydrogen sulfide or ammonia. In this second case, the chemosynthetic microorganisms are dependent on photosynthesis which occurs elsewhere and which produces the O_2 that they require.

Many chemosynthetic microorganisms are consumed by other organisms in the ocean, and symbiotic associations between chemosynthesizers and respiring heterotrophs are quite common. Large populations of animals can be supported by chemosynthetic primary production at hydrothermal vents, methane clathrates, cold seeps, and whale falls.

It has been hypothesized that chemosynthesis may support life below the surface of Mars, Jupiter's moon Europa, and other planets.

Hydrogen sulfide chemosynthesis: $CO_2 + O_2 + 4\{H_2S\}$? $CH_2O + 4\{S\} + 3\{H_2O\}$

USE OF TERM IN MOLECULAR NANOTECHNOLOGY

The term 'chemosynthesis' is also used in molecular nanotechnology broadly to refer to any chemical synthesis where reactions occur due to random thermal motion, a class which encompasses almost all of modern synthetic chemistry. The human-authored processes of chemical engineering are accordingly represented as biomimicry of the natural phenomena above, and the entire class of non-photosynthetic chains by which complex molecules are constructed is desribed as chemo.

This form of engineering is then contrasted with mechanosynthesis, a hypothetical process where individual molecules are mechanically manipulated to control reactions to human specification. Since photosynthesis and other natural processes create extremely complex molecules to the specifications contained in RNA and stored long-term in DNA form, advocates of molecular engineering claim that an artificial process can likewise exploit a chain of long-term storage, short-term storage, enzyme-like copying mechanisms similar to those in the cell, and ultimately produce complex molecules which need not be proteins. For instance, sheet diamond or carbon nanotubes could be produced by a chain of non-biological reactions that have been designed using the basic model of biology.

Use of the term chemosynthesis reinforces the view that this is feasible by pointing out that several alternate means of creating complex proteins, mineral shells of mollusks and crustaceans, etc., evolved naturally, not all of them dependent on photosynthesis and a food chain from the sun via chlorophyll. Since more than one such pathway exists to creating complex molecules, even extremely specific ones such as proteins edible to fish, the likelihood of humans being able to design an entirely new one is considered (by these advocates) to be near certainty in the long run, and possible within a generation.

Discovery

In 1890, Sergei Nikolaevich Vinogradskii (Winogradsky) proposed a novel life process called 'chemosynthesis'. His discovery suggested that some microbes could live solely on inorganic matter emerged during his physiological research in 1880s in Strassburg and Zurich on sulfur, iron, and nitrogen bacteria.

This was confirmed nearly 100 years later, when hydrothermal vents were predicted to exist in 1970s. The hot springs and strange creatures were discovered by Alvin, the world's first deep-sea submersible, expedition, in 1977 to the Galapagos Rift. Chemosynthesis told scientist that there could be another source of life besides sunlight.

A 2004 television mini-series hosted by Bill Nye named chemosynthesis as one of the 100 greatest scientific discoveries of all time.

CHEMOTROPH

Chemotrophs are organisms that obtain energy by the oxidation of electron donating molecules in their environments. These molecules can be organic (organotrophs) or inorganic (lithotrophs). The chemotroph designation is in contrast to phototrophs which utilize solar energy. Chemotrophs can be either autotrophic or heterotrophic.

- Chemoautotrophs (or chemotrophic autotroph), (*Gk: Chemo* = chemical, *auto* = self, *troph* = nourishment) in addition to deriving energy from chemical reactions, synthesize all necessary organic compounds from carbon dioxide. Chemoautotrophs generally only use inorganic energy sources. Most are bacteria or archaea that live in hostile environments such as deep sea vents and are the primary producers in such ecosystems. Evolutionary scientists believe that the first organisms to inhabit Earth were chemoautotrophs that produced oxygen as a by-product and later evolved into both aerobic, animal-like organisms and photosynthetic, plant-like organisms. Chemoautotrophs generally fall into several groups: methanogens, halophiles, sulfur reducers, nitrifiers, anammoxbacteria and thermoacidophiles.

- Chemoheterotrophs (or chemotrophic heterotrophs) (*Gk*: *Chemo* = chemical, *hetero* = (an)other, *troph* = nourishment) must ingest organic building blocks that they are incapable of creating on their own. Most chemoheterotrophs derive energy from organic molecules like glucose.

IRON AND MANGANESE OXIDIZING BACTERIA

In the deep oceans, iron oxidizing bacteria derive their energy needs by oxidizing Iron II to Iron III. The extra electron obtained from this reaction powers the cells, replacing or augmenting traditional phototrophism.

- In general, iron oxidizing bacteria can only exist in areas with high iron concentrations - such as new lava beds or areas of hydrothermal activity (where there is dissolved Fe). Most of the ocean is devoid of iron, due to both the oxidative effect of dissolved oxygen in the water and the tendency of prokaryotes to uptake the iron.
- Lava beds supply bacteria with iron straight from the earth's mantle, but only newly formed igneous rocks have high enough levels of unoxidized iron. In addition, since oxygen is necessary for the reaction, these bacteria are much more common in the upper ocean, where oxygen is more abundant.
- What is still unknown though is how exactly iron bacteria extract the iron out of the rock. It is accepted that some mechanism exists which eats away at the rock, perhaps through specialized enzymes or compounds which bring more FeO to the surface. It has been long debated about how much of the weathering of the rock is due to biotic and how much can be attributed to abiotic processes.
- Hydrothermal vents also release large quantities of dissolved iron into the deep ocean, allowing bacteria to survive. In addition, the high thermal gradient around vent systems means a wide variety of bacteria can coexist, each with its own specialized temperature niche.
- Regardless of the catalytic method used, chemoautotrophic bacteria provide a significant but frequently overlooked

food source for deep sea ecosystems - which otherwise receive limited sunlight and organic nutrients.

Manganese oxidizing bacteria also make use of igneous lava rocks in much the same way - by oxidizing Mn2+ into Mn4+. Manganese is much rarer than iron in oceanic crust, but is much easier for bacteria to extract from the igneous glass. In addition, each manganese oxidation yields around twice the energy as an iron oxidation due to the gain of twice the number of electrons. Much still remains unknown about manganese oxidizing bacteria because they have not been cultured and documented to any great extent.

6

Light-dependent Reaction

INTRODUCTION

The initial stage of the photosynthetic system is the light-dependent reaction, which converts solar energy into potential energy.

The light dependent reaction produces oxygen gas and converts ADP and $NADP^+$ into the energy carriers ATP and NADPH.

Light dependent reactions occur on the thylakoid membrane inside a chloroplast (this is where the chloroplasts are embedded). Inside the thylakoid membrane is called the thylakoid space, and outside the thylakoid membrane is the stroma.

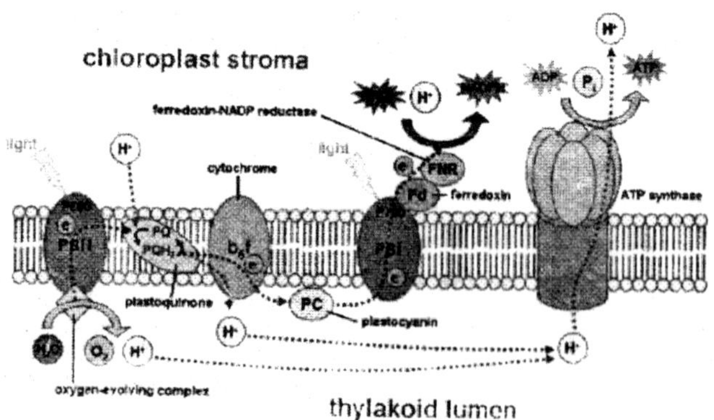

The chlorophyll's electron can follow either of two different pathways, cyclic or non-cyclic.

The first ideas about light being used in photosynthesis were proposed by Jan Ingenhousz in 1779 who recognized it was sunlight falling on plants that was required, although Joseph Priestly had noted the production of oxygen without the association with light in 1772. Cornelius Van Niel proposed in 1931 that photosynthesis is a case of general mechanism where a photon of light is used to photo decompose a hydrogen donor and the hydrogen being used to reduce CO_2. Then in 1939 Robin Hill showed that isolated chloroplasts would make oxygen, but not fix CO_2 showing the light and dark reactions occurred in different places . This led later to the discovery of photosystem 1 and 2.

Cyclic Photophosphorylation

In cyclic electron flow, the electron begins in a pigment complex called photosystem I, passes from the primary acceptor to ferredoxin, then to a complex of two cytochromes (similar to those found in mitochondria), and then to plastoquinone before returning to chlorophyll. This transport chain produces a proton-motive force, pumping H^+ ions across the membrane; this produces a concentration gradient which can be used to power ATP synthase during chemiosmosis. This pathway is known as cyclic photophosphorylation, and it produces O_2, as well as ATP. Unlike non-cyclic photophosphorylation, $NADP^+$ does not accept the electrons, but they are sent back to photosystem I. NADPH is NOT produced in cyclic photophosphorylation. In bacterial photosynthesis, a single photosystem is used, and therefore is involved in cyclic photophosphorylation.

NONCYCLIC PHOTOPHOSPHORYLATION

The other pathway, noncyclic photophosphorylation, is a two-stage process involving two different chlorophyll photosystems. Being a light reaction, Noncyclic photophosphorylation occurs on thylakoid membranes inside chloroplasts. First, a water molecule is broken down into $2H^+ + 1/2O_2 + 2e^-$ by a process called photolysis (or *light-splitting*). The two electrons from the water molecule are kept in photosystem II, while the $2H^+$ and $1/2O_2$ are left out for further use. Then a photon is absorbed by chlorophyll pigments on surrounding the reaction core center of the photosystem. The light excites the electrons of each pigment, causing a chain reaction which eventually transfers energy to the core of photosystem II, exciting the two electrons which are transferred to the primary electron acceptor. The deficit of electrons is replenished by taking electrons from another molecule of water. The electrons transfer from the primary acceptor to plastoquinone, then to plastocyanin, providing the energy for hydrogen ions (H^+) to be pumped into the thylakoid space. This creates a gradient, making H^+ ions flow back into the stroma of the chloroplast, providing the energy for the regeneration of ATP.

The photosystem II complex replaced its lost electrons from an external source, however, the two other electrons are not returned to photosystem II as they would in the analogous cyclic pathway. Instead, the still-excited electrons are transferred to a photosystem I complex, which boosts their energy level to a higher level using a second solar photon. The highly excited electrons are transferred to the acceptor molecule, but this time are passed on to an enzyme called Ferredoxin- NADP reductase $NADP^+$ reductase, for short FNR, which uses them to catalyst the reaction (as shown):

$$NADP^+ + 2H^+ + 2e^- ? NADPH + H^+$$

This consumes the H^+ ions produced by the splitting of water, leading to a net production of $1/2O_2$, ATP, and NADPH + H^+ with the consumption of solar photons and water.

The concentration of NADPH in the chloroplast may help regulate which pathway electrons take through the light reactions. When the chloroplast runs low on ATP for the Calvin cycle, NADPH will accumulate and the plant may shift from noncyclic to cyclic electron flow.

Steps

It is important to note that both photosystems are almost simultaneously excited; thus, both photosystems begin functioning at almost the same time.

The excited electron is passed along until it reaches P680 chlorophyll.

The excited electron is passed to the primary electron acceptor. Photolysis in the thylakoid takes the electrons from water and replaces the P_{680} electrons that were passed to the primary electron acceptor. (O_2 is released into the air as a waste product).

The electrons are passed to photosystem I via the electron transport chain (ETC) and in the process used to pump protons across the thylakoid membrane into the lumen.

The stored energy in the proton gradient is used to produce ATP which is used later in the Calvin-Benson Cycle.

P_{700} chlorophyll then uses light to excite the electron to its second primary acceptor.

The electron is sent down another ETC and used to reduce $NADP^+$ to NADPH.

The NADPH is then used later in the Calvin-Benson Cycle to remove PGA that is produced from RuBisCO reaction and releases enzyme for continuation of steady state reaction.

PHOTOSYNTHETIC REACTION CENTRE

A photosynthetic reaction center is a complex of three types of protein that is the site where molecular excitations originating from sunlight are transformed into a series of electron-transfer reactions. The reaction center proteins bind functional co-factors, chromophores or pigments such as chlorophyll and pheophytin molecules. These absorb light, promoting an electron to a higher energy level within a pigment. The free energy created is used to reduce a chain of electron acceptors which have subsequently lowered redox-potentials, and is critical for the production of chemical energy during photosynthesis.

Reaction centers are present in all green plants and in many bacteria and algae. Green plants have two reaction centers known as photosystem I and photosystem II and the structures of these centres are complex, involving a multisubunit protein. The reaction centre found in *Rhodopseudomonas* bacteria is currently better understood since it has fewer proteins than the examples in green plants.

Capturing Light Energy

A reaction centre is laid out in such a way that it captures the energy of a photon using pigment molecules and turns it into a usable form. Once the light energy has been absorbed directly by the pigment molecules, or passed to them by resonance transfer from surrounding antenna pigments, they release two electrons into an electron transport chain.

Light is made up of small bundles of energy called photons. If a photon with the right amount of energy hits an electron it will raise the electron to a higher energy level Electrons are most stable at their lowest energy level, what is also called its ground state. In this state the electron is in the orbit that has the least amount of energy. Electrons in higher energy levels can return to ground state in a manner analogous to a ball falling down a staircase. In doing so they release energy. This is the process which is exploited by a photosynthetic reaction centre.

When an electron rises to a higher energy level this increases the reduction potential of the molecule that the electron resides in. This means the molecule has a greater tendency to donate electrons, the key to the conversion of light energy to chemical energy. In green plants, the electron transport chain that follows has many electron acceptors including pheophytin, quinone, plastoquinone, cytochrome bf, and ferredoxin that ultimately result in the reduced molecule NADPH. The passage of the electron through the electron transport chain also results in the pumping of protons (hydrogen ions) from the chlorplast's stroma into the lumen resulting in a proton gradient across the thylakoid membrane that can be used to synthesis ATP using ATP synthase. Both the ATP and NADPH are used in the Calvin cycle to fix carbon dioxide into triose sugars.

BACTERIAL PHOTOSYNTHETIC REACTION CENTRE

The bacterial photosynthetic reaction centre has been an important model to understand the structure and chemistry of the biological process of capturing light energy. In the 1960s, Roderick Clayton was the first to purify the reaction centre complex from purple bacteria. However, the first crystal structure was determined in 1982 by Hartmut Michel, Johann Deisenhofer and Robert Huber for which they shared the Nobel Prize in 1988. This was also significant since it was the first structure for any membrane protein complex.

Four different subunits were found to be important for the function of the photosynthetic reaction centre. The L and M subunits, shown in blue and purple in the image of the structure both span the plasma membrane. They are structurally similar to one another, both having 5 transmembrane polypeptide helices Four bacteriochlorophyll b (BChl-b) molecules, two bacteriophaeophytin b molecules (BPh) molecules, two quinones (Q_A and Q_B), and a ferrous ion are associated with the L and M subunits. The H subunit, shown in gold, lies on the cytoplasmic side of the plasma membrane. A cytochrome subunit, shown in green, contains four c-type hemes and is located on the periplasmic surface (outer) of the membrane. The latter sub-unit is not a general structural motif in photosynthetic bacteria.

Reaction centres from different bacterial species may contain slightly altered bacterio-chlorophyll and bacterio-pheophytin chromophores as functional co-factors. These alterations causes shifts in the color of light that can be absorbed thus creating specific niches for photosynthesis. The reaction centre contains two pigments that serve to collect and transfer the energy from photon absorption: BChl and Bph. BChl roughly resembles the chlorophyll molecule found in green plants, but due to minor structural differences, its peak absorption wavelength is shifted into the infrared, with wavelengths as long as 1000nm. Bph has the same structure as BChl, but the central magnesium ion is replaced by two protons. This alteration causes both an absorbance maximum shift and a lowered redox-potential.

The Light Reaction

The process starts when light is absorbed by two BChl molecules that lie near the periplasmic side of the membrane. This pair of chlorophyll molecules, often called the "special pair", absorbs photons between 870nm and 960nm, depending on the species and thus is called P_{870} (for the species rhodobacter sphaeroides) or P_{960} (for rhodopseudomonas viridis), with *P* standing for "pair"). Once P absorbs a photon it ejects an electron, which is transferred through another molecule of Bchl to the BPh in the L subunit. This initial charge separation yields a positive charge on P and a negative charge on the BPh. This process takes place in 10 picoseconds (10^{-11} seconds).

The charges on the special pair $^+$ and the BPh^- could undergo charge recombination in this state. This would waste the high-energy electron and convert the absorbed light energy in to heat. Several factors of the reaction centre structure serve to prevent this. First the transfer of an electron from BPh^- to $P960^+$ is relatively slow compared to two other redox reactions in the reaction centre. The faster reactions involve the transfer of an electron from BPh^- (BPh^- is oxidised to BPh) to the electron acceptor quinone (Q_A) and the transfer of an electron to $P960^+$ ($P960^+$ is reduced to P960) from a heme in the cytochrome subunit above the reaction centre.

The high-energy electron which resides on the tightly bound quinone molecule Q_A is transferred to an exchangeable quinone molecule Q_B. This molecule is loosely associated with the protein and is fairly easy to detach. Two of the high-energy electrons are required to fully reduce Q_B to QH_2 taking up two protons from the cytoplasm in the process. The reduced quinone QH_2 diffuses through the membrane to another protein complex (cytochrome bc_1-complex) where it is oxidised. In the process the reducing power of the QH_2 is used to pump protons across the membrane to the periplasmic space. The electrons from the cytochrome bc_1-complex are then transferred through a soluble cytochrome c intermediate, called cytochrome c_2, in the periplasm to the cytochrome subunit. Thus, the flow of electrons in this system is cyclical.

OXYGENIC PHOTOSYNTHESIS

In 1772, the chemist Joseph Priestly carried out a series of experiments relating to the gasses involved in respiration and combustion. In his first experiment, he lit a candle and placed it under an upturned jar. After a short period of time, the candle burned out. He carried out a similar experiment with a mouse in the confined space of the burning candle. He found that the mouse died a short time after the candle had been extinguished. However, he could revivify the foul air by placing green plants in the area and exposing them to light. Priestly's observations were some of the first experiments that demonstrated the activity of a photosynthetic reaction centre.

In 1779, Jan Ingenhousz carried out more than 500 experiments spread out over 4 months in an attempt to understand what was really going on. He wrote up his discoveries in a book entitled '*Experiments upon Vegetables*'. Ingenhousz took green plants and immersed them in water inside a transparent tank. He observed many bubbles rising from the surface of the leaves whenever the plants were exposed to light. Ingenhousz collected the gas which was given off by the plants and performed several different tests in attempt to determine what the gas was. The test which finally revealed the identity of the gas was placing a smoldering taper into the gas sample and having it relight. This test proved it was oxygen, or as Joseph Priestly had called it, 'de-phlogisticated air'.

In 1932, Professor Robert Emerson and an undergraduate student, William Arnold, used a repetitive flash technique to precisely measure small quantities of oxygen evolved by chlorophyll in the algae *Chlorella*. Their experiment proved the existence of a photosynthetic unit. Gaffron and Wohl later interpreted the experiment and realized that the light absorbed by the photosynthetic unit was transferred. This reaction occurs at the reaction centre of photosystem II and takes place in cyanobacteria, algae and green plants.

CYANOBACTERIA PHOTOSYSTEM II, MONOMER, PDB 2AXT

Photosystem II is the photosystem that generates the electron that will eventually reduce $NADP^+$. Photosystem II is present on the thyakoid membranes inside chloroplasts, the site of photosynthesis in reen plants The structure of Photosystem II is remarkably simila to the bacterial reaction centre and it is theorized that they share a common ancestor.

The core of photosystem II consists of two subunits referred to as D_1 and D_2. These two subunits are similar to the L and M subunits present in the bacterial reaction centre. Photosystem II differs from the bacterial reaction centre in that it has many additional subunits which bind additional chlorophylls to increase efficiency. The overall reaction catalyzed by photosystem II is:

Q represents plastoquinone, the oxidized form of Q. QH_2 represents plastoquinol, the reduced form of Q. This process of reducing quinone is comparable to that which takes place in the bacterial reaction centre. Photosystem II obtains electrons by oxidizing water in a process called photolysis. Molecular oxygen is a byproduct of this process and it is this reaction that supplies the atmosphere with oxygen. The fact that the oxygen from green plants originated from water was first deduced by the Canadian-born American biochemist Martin David Kamen. He used a radioactive isotope of oxygen, O_{18} to trace the path of the oxygen, from water to gaseous molecular oxygen. This reaction is catalyzed by a reactive centre in photosystem II containing four manganese ions.

The reaction begins with the excitation of a pair of chlorophyll molecules similar to those in the bacterial reaction centre. Due to the presence of chlorophyll *a*, as opposed to bacteriochlorophyll, photosystem II absorbs light at a shorter wavelength. The pair of chlorophyll molecules at the reaction center are often referred to as P_{680}. When the photon has been absorbed the resulting high-energy electron is transferred to a nearby pheophytin molecule. The electron travels from the pheophytin molecule through two plastoquinone molecules, the first tightly bound, the second loosely bound. The tightly bound molecule is shown above the pheophytin molecule and is coloured red. The loosely bound molecule is to the left of this and is also coloured red. This flow of electrons is similar to that of the bacterial reaction centre. Two electrons are required to fully reduce the loosely bound plastoquinone molecule to QH_2 as well as the uptake of two protons.

The difference between photosystem II and the bacterial reaction centre is the source of the electron that neutralizes the pair of chlorophyll *a* molecules. In the bacterial reaction centre, the electron is obtained from a reduced compound heme group in a cytochrome subunit or from a water-soluble cytochrome-c protein.

A difference between photosystem II and the bacterial reaction centre is the source of the electron which neutralizes the pair of pigment molecules. Once photoinduced charge separation has taken place, the P_{680} molecule carries a positive charge. P_{680} is a very strong oxidant and extracts electrons from two water molecules which are bound at the manganese centre directly below the pair. This centre, below and to the left of the pair in the diagram, contains four manganese ions, a calcium ion, a chloride ion, and a tyrosine residue. Manganese is used because it is capable of existing in four oxidation states; Mn^{2+}, Mn^{3+}, Mn^{4+} and Mn^{5+}. Manganese also forms strong bonds with oxygen-containing molecules such as water.

Every time the P_{680} absorbs a photon, it emits an electron, gaining a positive charge. This charge is neutralized by the extraction of an electron from the manganese centre which sits directly below it. The process of oxidizing two molecules of water requires four electrons. The water molecules which are oxidized in the manganese centre are the source of the electrons which reduce the two molecules of Q to QH_2. To date, this water-splitting catalytic centre cannot be reproduced by any man-made catalyst.

Photosystem I

After the electron has left photosystem II it is transferred to a cytochrome b6f complex and then to plastocyanin, a blue copper protein and electron carrier. The plastocyanin complex carries the electron that will neutralize the pair in the next reaction centre, photosystem I.

As with photosystem II and the bacterial reaction centre, a pair of chlorophyll *a* molecules initiates photoinduced charge separation. This pair is referred to as P_{700}. 700 is a reference to the wavelength at which the chlorophyll molecules absorb light maximally. The P_{700} lies in the centre of the protein. Once photoinduced charge separation has been initiated, the electron travels down a pathway through a chlorophyll a molecule situated directly above the P_{700}, through a quinone molecule situated directly above that, through three $4F^{e}$-4S clusters and finally to an interchangeable ferredoxin complex. Ferredoxin is a soluble protein containing a $2F^{e}$-2S cluster coordinated by four cysteine residues. The positive charge left on the P_{700} is neutralized by the transfer of an electron from plastocyanin. Thus the overall reaction catalyzed by photosystem I is:

The cooperation between photosystems I and II creates an electron flow from H_2O to $NADP^+$. This pathway is called the 'Z-scheme' because the redox diagram from P_{680} to P_{700} resembles the letter z.

Chlorophyll

Chlorophyll is a green pigment found in most plants, algae, and cyanobacteria. Its name is derived from Greek: *chloros* "green") and *f* (*phyllon* "leaf"). Chlorophyll absorbs light most strongly in the blue and red but poorly in the green portions of the electromagnetic spectrum, hence the green colour of chlorophyll-containing tissues like plant leaves

CHLOROPHYLL AND PHOTOSYNTHESIS

Chlorophyll is vital for photosynthesis, which allows plants to obtain energy from light.

Chlorophyll molecules are specifically arranged in and around pigment protein complexes called photosystems which are embedded in the thylakoid membranes of chloroplasts. In these complexes, chlorophyll serves two primary functions. The function of the vast majority of chlorophyll (up to several hundred molecules per photosystem) is to absorb light and transfer that light energy by resonance energy transfer to a specific chlorophyll pair in the reaction center of the photosystems. Because of chlorophyll's selectivity regarding the wavelength of light it absorbs, areas of a leaf containing the molecule will appear green.

There are currently two accepted photosystem units, Photosystem II and Photosystem I, which have their own distinct reaction center chlorophylls, named P_{680} and P_{700}, respectively. These pigments are named after the wavelength (in nanometers) of their red-peak absorption maximum. The identity, function and spectral properties of the types of chlorophyll in each photosystem are distinct and determined by each other and the protein structure surrounding them. Once extracted from the protein into a solvent (such as acetone or methanol these chlorophyll pigments can be separated in a simple paper chromatography experiment, and, based on thc number of polar groups between chlorophyll a and chlorophyll b, will chemically separate out on the paper.

The function of the reaction center chlorophyll is to use the energy absorbed by and transferred to it from the other chlorophyll pigments in the photosystems to undergo a charge separation, a specific redox reaction in which the chlorophyll donates an electron into a series of molecular intermediates called an electron transport chain. The charged reaction center chlorophyll (P_{680}^+) is then reduced back to its ground state by accepting an electron. In Photosystem II, the electron which reduces P_{680}^+ ultimately comes from the oxidation of water into O_2 and H^+ through several intermediates. This reaction is how photosynthetic organisms like plants produce O_2 gas, and is the source for practically all the O_2 in Earth's atmosphere. Photosystem I typically works in series with Photosystem II, thus the P_{700}^+ of Photosystem I is usually reduced, via many intermediates in the thylakoid membrane, by electrons ultimately from Photosystem II. Electron transfer reactions in the thylakoid membranes are complex, however, and the source of electrons used to reduce P_{700}^+ can vary.

The electron flow produced by the reaction center chlorophyll pigments is used to shuttle H^+ ions across the thylakoid membrane, setting up a chemiosmotic potential mainly used to produce ATP chemical energy, and those electrons ultimately reduce $NADP^+$ to NADPH a universal reductant used to reduce CO_2 into sugars as well as for other biosynthetic reductions.

Reaction center chlorophyll-protein complexes are capable of directly absorbing light and performing charge separation events without other chlorophyll pigments, but the absorption cross section (the likelihood of absorbing a photon under a given light intensity) is small. Thus, the remaining chlorophylls in the photosystem and antenna pigment protein complexes associated with the photosystems all cooperatively absorb and funnel light energy to the reaction center. Besides chlorophyll *a*, there are other pigments, called accessory pigments, which occur in these pigment-protein antenna complexes.

Chemical Structure

Chlorophyll is a chlorin pigment, which is structurally similar to and produced through the same metabolic pathway as other porphyrin pigments such as heme. At the center of the chlorin ring is a magnesium ion. The chlorin ring can have several different side chains, usually including a long phytol chain. There are a few different forms that occur naturally, but the most widely distributed form in terrestrial plants is chlorophyll *a*. The general structure of chlorophyll *a* was elucidated by Hans Fischer in 1940, and by 1960, when most of the stereochemistry of chlorophyll *a* was known, Robert Burns Woodward published a total synthesis of the molecule as then known. In 1967, the last remaining stereochemical elucidation was completed by Ian Fleming, and in 1990 Woodward and co-authors published an updated synthesis

Spectrophotometry

Absorbance spectra of free chlorophyll *a* (green) and *b* (red) in a solvent. The spectra of chlorophyll molecules are slightly modified *in vivo* depending on specific pigment-protein interactions.

Measurement of the absorption of light is complicated by the solvent used to extract it from plant material, which affects the values obtained.

- In diethyl ether, chlorophyll a has approximate absorbance maxima of 430 nm and 662 nm, while chlorophyll b has approximate maxima of 453 nm and 642 nm.

- The absorption peaks of Chlorophyll a are at 665 nm and 465 nm. Chlorophyll a fluoresces at 673 nm. The peak molar absorption coefficient of chlorophyll a exceeds 10^5 M^{-1} cm^{-1}, which is among the highest for organic compounds.

Biosynthesis

In plants, chlorophyll may be synthesized from succinyl-CoA and glycine, although the immediate precursor to chlorophyll *a* and *b* is protochlorophyll.

Chlorosis is a condition in which leaves produce insufficient chlorophyll, turning them yellow. Chlorosis can be caused by a nutrient deficiency including iron - called iron chlorosis, or in a shortage of magnesium or nitrogen. Soil pH sometimes play a role in nutrient-caused chlorosis, many plants are adapted to grow in soils with specific pHs and their ability to absorb nutrients from the soil can be dependent on the soil pH Chlorosis can also be caused by pathogens including viruses, bacteria and fungal infections or sap sucking insects.

Culinary Use

Chefs use chlorophyll to colour a variety of food and beverages green, such as pasta, or absinthe. Chlorophyll is not soluble in water and is first mixed with a small quantity of oil to obtain the desired result.

7

Circadian Rhythm

INTRODUCTION

A circadian rhythm is an approximate daily periodicity, a roughly-24-hour cycle in the biochemical, physiological or behavioural processes of living beings, including plants, animals, fungi and cyanobacteria. The term "circadian", coined by Franz Halberg, comes from the Latin *circa*, "around", and *diem* or *dies*, "day", meaning literally "approximately one day." The formal study of biological temporal rhythms such as daily, tidal, weekly, seasonal, and annual rhythms, is called chronobiology.

Circadian rhythms are endogenously generated, and can be entrained by external cues, called Zeitgebers. The primary one is daylight. These rhythms allow organisms to anticipate and prepare for precise and regular environmental changes.

The earliest known account of a circadian rhythm dates from the 4th century BC, when Androsthenes, in descriptions of the marches of Alexander the Great, described diurnal leaf movements of the tamarind tree. The first modern observation of endogenous circadian oscillation was by the French scientist Jean-Jacques d'Ortous de Mairan in the 1700s; he noted that 24-hour patterns in the movement of the leaves of the plant *Mimosa pudica* continued even when the plants were isolated from external stimuli.

In 1918 J.S. Szymanski showed that animals are capable of maintaining 24-hour activity patterns in the absence of external cues such as light and changes in temperature. It is also recorded that the Circadian rhythm is the main cycle that is linked with the hypothalamus.

CRITERIA

Three general criteria of circadian rhythms are necessary to differentiate genuinely endogenous rhythms from coincidental or apparent ones: the rhythms persist in the absence of cues, they can be brought to match the local time, and will do so in a precise manner over a range of temperatures:

- The rhythm persists in constant conditions (for example, constant dark) with a period of about 24 hours. The rationale for this criterion is to distinguish circadian rhythms from those "apparent" rhythms which merely are responses to external periodic cues. A rhythm cannot be declared to be endogenous unless it has been tested in conditions without external periodic input.
- The rhythm is temperature-compensated, i.e. it maintains the same period over a range of temperatures. The rationale for this criterion is to distinguish circadian rhythms from other biological rhythms arising due to the circular nature of a reaction pathway. At a low enough or high enough temperature, the period of a circular reaction may reach 24 hours, but it will be merely coincidental.
- The rhythm can be reset by exposure to an external stimulus. The rationale for this criterion is to distinguish circadian rhythms from other imaginable endogenous 24-hour rhythms that are immune to resetting by external cues and hence do not serve the purpose of estimating the local time. Travel across time zones illustrates the necessity of the ability to adjust the biological clock so that it can reflect the local time and anticipate what will happen next.

Photosensitive proteins and circadian rhythms are believed to have originated in the earliest cells, with the purpose of protecting the replicating of DNA from high ultraviolet radiation during the daytime. As a result, replication was relegated to the dark. The fungus *Neurospora,* which exists today, retains this clock-regulated mechanism. Rhythmicity appears to be as important in regulating cyclic biochemical processes within an individual, as in coordinating with the environment. This is suggested by the maintenance (heritability) of circadian rhythms in fruit flies after several hundred generations in constant laboratory conditions, as well as the experimental elimination of behavioral but not physiological circadian rhythms in quail.

The simplest known circadian clock is that of the prokaryotic cyanobacteria. Recent research has demonstrated that the circadian clock of *Synechococcus elongatus* can be reconstituted *in vitro* with just the three proteins of their central oscillator. This clock has been shown to sustain a 22-hour rhythm over several days upon the addition of ATP. Previous explanations of the prokaryotic circadian timekeeper were dependent upon a DNA transcription/translation feedback mechanism. It is an outstanding question whether circadian clocks in eukaryotic organisms require translation/transcription-derived oscillations. For although the circadian systems of eukaryotes and prokaryotes have the same basic architecture: input-central oscillator-output, they do not share any homology. This implies probable independent origins.

In 1971, Ronald J. Konopka and Seymour Benzer first identified a genetic component of the biological clock using the fruit fly as a model system. Three mutant lines of flies displayed aberrant behaviour - one had a shorter period, another had a longer one and the third had none. All three mutations mapped to the same gene, which was named *period* The same gene was identified to be defective in the sleep disorder FASPS (Familial Advanced Sleep Phase Syndrome) in human beings thirty years later - underscoring the conserved nature of the molecular circadian clock through evolution. We now know many more genetic components of the biological clock. Their interactions result in an interlocked feedback loop of gene products resulting in periodic fluctuations that the cells of the body interpret as a specific time of the day.

A great deal of research on biological clocks was done in the latter half of the 20th century. It is now known that the molecular circadian clock can function within a single cell, i.e., it is cell-autonomous. At the same time, different cells may communicate with each other resulting in a synchronised output of electrical signaling. These may interface with endocrine glands of the brain to result in periodic release of hormones. The receptors for these hormones may be located far across the body and synchronise the peripheral clocks of various organs. Thus, the information of the time of the day as relayed by the eyes travels to the clock in the brain, and, through that, clocks in the rest of the body may be synchronised. This is how the timing of, for example, sleep/wake, body temperature, thirst, and appetite are coordinately controlled by the biological clock.

IMPORTANCE IN ANIMALS

Circadian rhythms are important in determining the sleeping and feeding patterns of all animals, including human beings. There are clear patterns of core body temperature, brain wave activity, hormone production, cell regeneration and other biological activities linked to this daily cycle. In addition, photoperiodism, the physiological reaction of organisms to the length of day or night, is vital to both plants and animals, and the circadian system plays a role in the measurement and interpretation of daylength.

Timely prediction of seasonal periods of weather conditions, food availability or predator activity is crucial for survival of many species. Although not the only parameter, the changing length of the photoperiod ('daylength') is the most predictive environmental cue for the seasonal timing of physiology and behavior, most notably for timing of migration, hibernation and reproduction.

Impact of Light-dark Cycle

The rhythm is linked to the light-dark cycle. Animals, including humans, kept in total darkness for extended periods eventually function with a freerunning rhythm. Each "day," their sleep cycle is pushed back or forward, depending on whether their endogenous period is shorter or longer than 24 hours. The environmental cues that each day reset the rhythms are called *Zeitgebers* (from the German, *Time Givers*) It is interesting to note that totally-blind subterranean mammals (e.g., blind mole rat *Spalax* sp.) are able to maintain their endogenous clocks in the apparent absence of external stimuli.

Freerunning organisms that normally have one consolidated sleep episode will still have it when in an environment shielded from external cues, but the rhythm is, of course, not entrained to the 24-hour light/dark cycle in nature. The sleep/wake rhythm may, in these circumstances, become out of phase with other circadian or ultradian rhythms such as temperature and digestion

Recent research has influenced the design of spacecraft environments, as systems that mimic the light/dark cycle have been found to be highly beneficial to astronauts.

Arctic Animals

Norwegian researchers at the University of Tromsø have shown that some Arctic animals (ptarmigan, reindeer) show circadian rhythms only in the parts of the year that have daily sunrises and sunsets. In one study of reindeer, animals at 70 degrees North showed circadian rhythms in the autumn, winter, and spring, but not in the summer. Reindeer at 78 degrees North showed such rhythms only autumn and spring. The researchers suspect that other Arctic animals as well may not show circadian rhythms in the constant light of summer and the constant dark of winter.

However, another study in northern Alaska found that ground squirrels and porcupines strictly maintained their circadian rhythms through 82 days and nights of sunshine. The researchers speculate that these two small mammals see that the apparent distance between the sun and the horizon is shortest once a day, and, thus, a sufficient signal to adjust by.

THE BIOLOGICAL CLOCK IN MAMMALS

The primary circadian "clock" in mammals is located in the suprachiasmatic nucleus (or nuclei) (SCN), a pair of distinct groups of cells located in the hypothalamus. Destruction of the SCN results in the complete absence of a regular sleep/wake rhythm. The SCN receives information about illumination through the eyes. The retina of the eyes contains not only "classical" photoreceptors but also photoresponsive retinal ganglion cells. These cells, which contain a photo pigment called melanopsin, follow a pathway called the retinohypothalamic tract, leading to the SCN. If cells from the SCN are removed and cultured, they maintain their own rhythm in the absence of external cues.

It appears that the SCN takes the information on day length from the retina, interprets it, and passes it on to the pineal gland, a tiny structure shaped like a pine cone and located on the epithalamus. In response the pineal secretes the hormone melatonin. Secretion of melatonin peaks at night and ebbs during the day.

The circadian rhythms of humans can be entrained to slightly shorter and longer periods than the earth's 24 hours. Researchers at Harvard have recently shown that human subjects can at least be entrained to a 23.5-hour cycle and a 24.65-hour cycle (the latter being the natural solar day-night cycle on the planet Mars).

Determining the Human Circadian Rhythm

The classic phase markers for measuring the timing of a mammal's circadian rhythm are melatonin secretion by the pineal gland and core body temperature.

For temperature studies, people must remain awake but calm and semi-reclined in near darkness while their rectal temperatures are taken continuously. The average human adult's temperature reaches its minimum at about 05:00 (5 a.m.), about two hours before habitual wake time, though variation is great among normal chronotypes.

Melatonin is absent from the system or undetectably low during daytime. Its onset in dim light, *dim-light melatonin onset* (DLMO), at about 21:00 (9 p.m.) can be measured in the blood or the saliva. Both DLMO and the midpoint (in time) of the presence of the hormone in the blood or saliva have been used as circadian markers.

However, newer research indicates that the melatonin *offset* may be the most reliable marker. In 2005 found that melatonin phase markers were more stable and more highly correlated with the timing of sleep than the core temperature minimum. Both sleep offset and melatonin offset were more strongly correlated with the various phase markers than sleep onset. In addition, the declining phase of the melatonin levels was more reliable and stable than the termination of melatonin synthesis.

One method used for measuring melatonin offset is to analyze a sequence of urine samples throughout the morning for the presence of the melatonin metabolite 6-sulphatoxy-melatonin (aMT6s). In a study which confirmed the frequently found delayed circadian phase in healthy adolescents.

Outside the "Master Clock"

More-or-less independent circadian rhythms are found in many organs and cells in the body outside the suprachiasmatic nuclei (SCN), the "master clock." These clocks, called peripheral oscillators, are found in the esophagus, lung, liver, pancreas, spleen, thymus and the skin. Though oscillators in the skin respond to light, a systemic influence has not been proven so far. There is some evidence that also the olfactory bulb and prostate may experience oscillations when cultured, suggesting that also these structures may be weak oscillators.

Furthermore, liver cells, for example, appear to respond to feeding rather than to light. Cells from many parts of the body appear to have freerunning rhythms.

Light and the Biological Clock

Light resets the biological clock in accordance with the phase response curve (PRC). Depending on the timing, light can advance or delay the circadian rhythm. Both the PRC and the required illuminance vary from species to species; much lower light levels are required to reset the clocks in nocturnal rodents than in humans.

In addition to light intensity, wavelength (or colour) of light is an important factor in the degree to which the clock is reset.

Melanopsin is most efficiently excited by blue light, 420-440 nm according to some researchers while others have reported 470-485nm.

The Myth of the 25-hour Day

Early investigators determined the human circadian period to be 25 hours or more. They went to great lengths to shield subjects from time cues and daylight, but they were not aware of the effects of indoor electric lights. The subjects were allowed to turn on light when they were awake and to turn it off when they wanted to sleep. Electric light in the evening delayed their circadian phase. These results became well known.

The Human Circadian Period

Modern research under very controlled conditions has shown the human period for adults to be just slightly longer than 24 hours on average. The range for normal, healthy adults of all ages to be quite narrow: 24 hours and 11 minutes ± 16 minutes. The "clock" resets itself daily to the 24-hour cycle of the earth's rotation.

Human Health

Timing of medical treatment in coordination with the body clock may significantly increase efficacy and reduce drug toxicity or adverse reactions. For example, appropriately timed treatment with angiotensin converting enzyme inhibitors (ACEi) may reduce nocturnal blood pressure and also benefit left ventricular (reverse) remodeling. A number of studies have also concluded that a short period of sleep during the day (commonly referred to as a power-nap) does not have any affect on normal circadian rhythm, yet a power-nap can decrease stress and improve productivity.

There are many health problems associated with a disturbance in the human circadian rhythm, such as Seasonal Affective Disorder (SAD), delayed sleep phase syndrome (DSPS) and other circadian rhythm disorders. Circadian rhythms also play a part in the reticular activating system which is crucial for maintaining a state of consciousness. In addition, a reversal in the sleep-wake cycle may be a sign or complication of uremia, azotemia or acute renal failure.

Disruption

Disruption to rhythms usually has a negative effect. Many travelers have experienced the condition known as jet lag, with its associated symptoms of fatigue, disorientation and insomnia.

A number of other disorders, for example bipolar disorder and some sleep disorders, are associated with irregular or pathological functioning of circadian rhythms. Recent research suggests that circadian rhythm disturbances found in bipolar disorder are positively influenced by lithium's effect on clock genes.

Disruption to rhythms in the longer term is believed to have significant adverse health consequences on peripheral organs outside the brain, particularly in the development or exacerbation of cardiovascular disease. The suppression of melatonin production associated with the disruption of the circadian rhythm may increase the risk of developing cancer.

Effects on Cocaine Sensitization in Mice

Circadian rhythms and clock genes expressed in brain regions outside the SCN may significantly influence the effects produced by drugs such as cocaine. Moreover, genetic manipulations of clock genes profoundly affect cocaine's actions.

LIGHT EFFECTS ON CIRCADIAN RHYTHM

Numerous organisms maintain inherent individual rhythms to biological processes, known as circadian rhythms, that assist the organism in maintaining functional periodicity relative to the 24 hour day/night cycle of the earth. These rhythms are maintained by the individual organisms, but due to variable individuality and environmental pressures, must be continually reset to synch with the natural environmental cycle. In order for this to be accomplished, external factors must play some role in the synchronization, or entrainment, of the internal circadian rhythm with the external environment. Of the various factors that influence this entrainment, light exposure to the eyes is the strongest effecter.

Demonstrated Effects

All of the mechanisms of light-effected entrainment are not yet fully known, however numerous studies have demonstrated the effectiveness of light entrainment to the day/night cycle. Studies have shown that:

- The time of exposure to light influences entrainment; see phase response curve.
  - Exposure to bright light after wakening advances the circadian rhythm, whereas exposure before sleeping delays the rhythm.
- The length of exposure influences entrainment.
  - Longer exposures have a greater effect than shorter exposures.
  - Consistent exposure has a greater effect than intermittent exposure.
  - Constant exposure eventually disrupts the cycle to the point that other functions like memory and stress coping may be impaired.
- Intensity and wavelength of light influence entrainment.
  - Brighter light is more effective than dim light.
  - Dim light can effect entrainment relative to darkness.
  - Low intensity short wavelength (blue) light may be equally effective as high intensity white light.

Internal Regulators

Light's effect on the circadian rhythms of all or most animals has been well-documented. However, since circadian rhythms are internal functions, the influence of external factors like light and an individual's sensitivity to them can to some degree be regulated by internal mechanisms.

- For example, evidence of a negative regulation of light-dependent gene transcription has been found in zebrafish. In this study overabundance of the enzyme Catalase reduced the transcription of genes that were dependent on light, whereas inhibition of the enzyme resulted in increased transcription.

- Another study found that a deficit of the oligopeptide angiotensin in the brain of laboratory rats resulted in delayed adjustment to changes in the day/night pattern.
- Similarly, deficits of TrkB tyrosine kinase in mice, a receptor for brain-derived neurotrophic factor (BDNF), result in a decrease of the ability to entrain to shifts in the day/night cycle.

Internal conditions may thus sway the effectiveness of entrainment to light. All mechanisms behind the process are not yet fully understood.

Other Factors

Although many researchers consider light to be the strongest cue for entrainment, it is by no means the only factor acting on circadian rhythms. Other factors may enhance or detract the effectiveness of entrainment. For instance, physical activity like exercise when coupled with light exposure results in a somewhat stronger entrainment response. Other factors such as music and administration of the neurohormone melatonin have shown similar effects. Numerous other factors affect entrainment as well. Temperature, pharmacology, locomotor stimuli, social interaction, sexual stimuli, stress, and many others have also been shown to effect circadian entrainment, even in absence of cues from light.

8

Photochromism

INTRODUCTION

Photochromism is the reversible transformation of a chemical species between two forms by the absorption of electromagnetic radiation, where the two forms have different absorption spectra.[Trivially, this can be described as a reversible change of color upon exposure to light. The phenomenon was discovered in the late 1880s, including work by Markwald, who studied the reversible change of color of 2, 3, 4, 4-tetrachloronaphthalen-1(4H)-one in the solid state. He labeled this phenomenon "phototropy", and this name was used until the 1950s when Yehuda Hirshberg, of the Weizmann Institute of Science in Israel proposed the term "photochromism". Photochromism can take place in both organic and inorganic compounds, and also has its place in biological systems (for example retinal in the vision process).

Photochromism does not have a rigorous definition, but is usually used to describe compounds that undergo a reversible photochemical reaction where an absorption band in the visible part of the electromagnetic spectrum changes dramatically in strength or wavelength. In many cases, an absorbance band is present in only one form. The degree of change required for a photochemical reaction to be dubbed "photochromic" is that which appears dramatic by eye, but in essence there is no dividing line between photochromic reactions and other photochemistry. Therefore, while the trans-cis isomerization of azobenzene is considered a photochromic reaction, the analogous reaction of stilbene is not. Since photochromism is just a special case of a photochemical reaction, almost any photochemical reaction type may be used to produce photochromism with appropriate molecular design. Some of the most common processes involved in photochromism are pericyclic reactions, cis-trans isomerizations, intramolecular hydrogen transfer, intramolecular group transfers, dissociation processes and electron transfers (oxidation-reduction).

Another somewhat arbitrary requirement of photochromism is that it requires the two states of the molecule to be thermally stable under ambient conditions for a reasonable time. All the same, nitrospiropyran (which back-isomerizes in the dark over ~10 minutes at room temperature) is considered photochromic. All photochromic molecules back-isomerize to their more stable form at some rate, and this back-isomerization is accelerated by heating. There is therefore a close relationship between photochromic and thermochromic compounds. The timescale of thermal back-isomerization is important for applications, and may be molecularly engineered. Photochromic compounds considered to be "thermally stable" include some diarylethenes, which do not back isomerize even after heating at 80° C for 3 months.

Since photochromic chromophores are dyes, and operate according to well-known reactions, their molecular engineering to fine-tune their properties can be achieved relatively easily using known design models, quantum mechanics calculations, and experimentation. In particular, the tuning of absorbance bands to particular parts of the spectrum and the engineering of thermal stability have received much attention.

Sometimes, and particularly in the dye industry, the term "irreversible photochromic" is used to describe materials that undergo a permanent color change upon exposure to ultraviolet or visible light radiation. Because by definition photochromics are reversible, there is technically no such thing as an "irreversible photochromic"—this is loose usage, and these compounds are better referred to as "photochangable" or "photoreactive" dyes.

Apart from the qualities already mentioned, several other properties of photochromics are important for their use. These include:

- **Quantum yield** of the photochemical reaction. This determined the efficiency of the photochromic change with respect to the amount of light absorbed. The quantum yield of isomerization can be strongly dependent on conditions.
- **Fatigue resistance**. In photochromic materials, fatigue refers to the loss of reversibility by processes such as photodegradation, photobleaching, photooxidation, and other side reactions. All photochromics suffer fatigue to some extent, and its rate is strongly dependent on the activating light and the conditions of the sample.
- **Photostationary state**. Photochromic materials have two states, and their interconversion can be controlled using different wavelengths of light. Excitation with any given wavelength of light will result in a mixture of the two states at a particular ratio, called the "photostationary state". In a perfect system, there would exist wavelengths that can be used to provide 1:0 and 0:1 ratios of the isomers, but in real systems this is not possible, since the active absorbance bands always overlap to some extent.
- **Polarity and solubility**. In order to incorporate photochromics in working systems, they suffer the same issues as other dyes. They are often charged in one or more state, leading to very high polarity and possible large changes in polarity. They also often contain large conjugated systems that limit their solubility.

PHOTOCHROMIC COMPLEXES

A *photochromic complex* is a kind of chemical compound that has photoresponsive parts on its ligand. These complexes have a specific structure: photoswitchable organic compounds are attached to metal complexes. For the photocontrollable parts, thermally and photochemically stable chromophores (azobenzene, diarylethene, spiropyran, etc.) are usually used. And for the metal complexes, a wide variety of compounds that have various functions (redox response, luminescence, magnetism, etc.) are applied.

The photochromic parts and metal parts are so close that they can affect each other's molecular orbitals. The physical properties of these compounds shown by parts of them (i.e., chromophores or metals) thus can be controlled by switching their other sites by external stimuli. For example, photoisomerization behaviors of some complexes can be switched by oxidation and reduction of their metal parts. Some other compounds can be changed in their luminescence behavior, magnetic interaction of metal sites, or stability of metal-to-ligand coordination by photoisomerization of their photochromic parts.

CLASSES OF PHOTOCHROMIC MATERIALS

Photochromic molecules can belong to various classes: triarylmethanes, stilbenes, azastilbenes, nitrones, fulgides, spiropyrans, naphthopyrans, spiro-oxazines, quinones and others.

Spiropyrans and Sprioxazines

Spiro-mero Photochromism

One of the oldest, and perhaps the most studied, families of photochromes are the spiropyrans. Very closely related to these are the spirooxazines. For example, the spiro form of an oxazine is a colorless leuco dye; the conjugated system of the oxazine and another aromatic part of the molecule is separated by a sp^3-hybridized "spiro" carbon. After irradiation with UV light, the bond between the spiro-carbon and the oxazine breaks, the ring opens, the spiro carbon achieves sp^2 hybridization and becomes planar, the aromatic group rotates, aligns its p-orbitals with the rest of the molecule, and a conjugated system forms with ability to absorb photons of visible light, and therefore appear colorful. When the UV source is removed, the molecules gradually relax to their ground state, the carbon-oxygen bond reforms, the spiro-carbon becomes sp^3 hybridized again, and the molecule returns to its colourless state.

This class of photochromes in particular are thermodynamically unstable in one form and revert to the stable form in the dark unless cooled to low temperatures. Their lifetime can also be affected by exposure to UV light. Like most organic dyes they are susceptible to degradation by oxygen and free radicals. Incorporation of the dyes into a polymer matrix, adding a stabilizer, or providing a barrier to oxygen and chemicals by other means prolongs their lifetime.

Dithienylethene Photochemistry

The "diarylethenes" were first introduced by Irie and have since gained widespread interest, largely on account of their high thermodynamic stability. They operate by means of a 6-pi electrocyclic reaction, the thermal analog of which is impossible due to steric hindrance. Pure photochromic dyes usually have the appearance of a crystalline powder, and in order to achieve the color change, they usually have to be dissolved in a solvent or dispersed in a suitable matrix. However, some diarylethenes have so little shape change upon isomerization that they can be converted while remaining in crystalline form.

Azobenzene Photoisomerization

The photochromic trans-cis isomerization of azobenzenes has been used extensively in molecular switches, often taking advantage of its shape change upon isomerization to produce a supramolecular result. In particular, azobenzenes incorporated into crown ethers give switchable receptors and azobenzenes in monolayers can provide light-controlled changes in surface properties.

PHOTOCHROMIC QUINONES

Some quinones, and phenoxynaphthacene quinone in particular, have photochromicity resulting from the ability of the phenyl group to migrate from one oxygen atome to another. Quinones with good thermal stability have been prepared, and they also have the additional feature of redox activity, leading to the construction of many-state molecular switches that operate by a mixture of photonic and electronic stimuli.

Inorganic Photochromics

Many inorganic substances also exhibit photochromic properties, often with much better resistance to fatigue than organic photochromics. In particular, silver chloride is extensively used in the manufacture of photochromic lenses. Other silver and zinc halides are also photochromic.

Applications

Sunglasses

One of the most famous reversible photochromic applications is color changing lenses for sunglasses, as found in eye-glasses. The largest limitation in using PC technology is that the materials cannot be made stable enough to withstand thousands of hours of outdoor exposure so long-term outdoor applications are not appropriate at this time.

The switching speed of photochromic dyes is highly sensitive to the rigidity of the environment around the dye. As result, they switch most rapidly in solution and slowest in the rigid environment like a polymer lens. Recently it has been reported that attaching flexible, low Tg polymers [(for example siloxanes or poly(butyl acrylate)] to the dyes allows them to switch much more rapidly in a rigid lens. Some spirooxazines with siloxane polymers attached switch at near solution-like speeds even though they are in a rigid lens matrix.

Supramolecular Chemistry

Photochromic units have been employed extensively in supramolecular chemistry. Their ability to give a light-controlled reversible shape change means that they can be used to make or break molecular recognition motifs, or to cause a consequent shape change in their surroundings. Thus, photochromic units have been demonstrated as components of molecular switches. The coupling of photochromic units to enzymes or enzyme cofactors even provides the ability to reversibly turn enzymes "on" and "off", by altering their shape or orientation in such a way that their functions are either "working" or "broken".

Data Storage

The possibility of using photochromic compounds for data storage was first suggested in 1956 by Yehuda Hirshberg. Since that time, there have been many investigations by various academic and commercial groups, particularly in the area of 3D optical data storage which promises discs that can hold a terabyte of data. Initially, issues with thermal back-reactions and destructive reading dogged these studies, but more recently more-stable systems have been developed.

Reversible photochromics are also found in applications such as toys, cosmetics, clothing and industrial applications. If necessary, they can be made to change between desired colors by combination with a permanent pigment.

PHOTOSYSTEM

Photosystems (ancient Greek: *phos* = light and *systema* = assembly) are protein complexes involved in photosynthesis. They are found in the thylakoid membranes of plants, algae and cyanobacteria (in plants and algae these are located in the chloroplasts), or in the cytoplasmic membrane of photosynthetic bacteria. A photosystem (or Reaction Center) is an enzyme which uses light to reduce molecules. This membrane protein complex is made of several subunits and contains numerous cofactors. In the photosynthetic membranes, reaction centers provide the driving force for the bioenergetic electron and proton transfer chain. When light is absorbed by a reaction center (either directly or passed by neighbouring pigment-antennae), a series of oxido-reduction reactions is initiated, leading to the reduction of a terminal acceptor. Two families of photosystems exist: type I reaction centers (like photosystem I (P_{700}) in chloroplasts and in green-sulphur bacteria) and type II reaction centers (like photosystem II (P_{680}) in chloroplasts and in non-sulphur purple bacteria). Each photosystem can be identified by the wavelength of light to which it is most reactive (700 and 680 nanometers, respectively for PSI and PSII in chloroplasts), and the type of terminal electron acceptor. Type I photosystems use ferredoxin-like iron-sulfur cluster proteins as terminal electron acceptors, while type II photosystems ultimately shuttle electrons to a quinone terminal electron acceptor. One has to note that both reaction center types are present in chloroplasts and cyanobacteria, working together to form a unique photosynthetic chain able to extract electrons from water, creating oxygen as a byproduct.

Structure

A reaction center comprises several (>10 or >11) protein subunits, providing a scaffold for a series of cofactors. The latter can be pigments (like chlorophyll, pheophytin, carotenoids), quinones or iron-sulfur clusters. Because chlorophyll *a* can only absorb light of a narrow wavelength, it works with the antenna pigments to gain energy from a larger part of the spectrum. The pigments absorb light of various wavelengths and pass along their gained energy to the reaction center chlorophyll. When the energy reaches the chlorophyll *a*, it releases two electrons into an electron transport chain.

Though chlorophyll *a* normally has an optimal absorption wavelength of 660 nanometers, it associates with different proteins in each type of photosystem to slightly shift its optimal wavelength, producing two distinct photosystem types. Other proteins serve to support the structure and electron pathways in the photosystem.

RELATIONSHIP BETWEEN PHOTOSYSTEMS I AND II

Historically photosystem I was named "I" since it was discovered before photosystem II, but this does not represent the order of the electron flow.

When photosystem II absorbs light, electrons in the reaction-center chlorophyll are excited to a higher energy level and are trapped by the primary electron acceptors. To replenish the deficit of electrons, electrons are extracted from water (either through photolysis or enzymatic means) and supplied to the chlorophyll.

Photoexcited electrons travel through the cytochrome b6f complex to photosystem I via an electron transport chain set in the thylakoid membrane. This energy fall is harnessed, (the whole process termed chemiosmosis), to transport hydrogen (H+)

through the membrane to provide a proton-motive force to generate ATP. If electrons only pass through once, the process is termed noncyclic photophosphorylation.

When the electron reaches photosystem I, it fills the electron deficit of the reaction-center chlorophyll of photosystem I. The deficit is due to photo-excitation of electrons which are again trapped in an electron acceptor molecule, this time that of photosystem I. Photosystems are z shaped.

These electrons may either continue to go through cyclic electron transport around PS I, or pass, via ferredoxin, to the enzyme NADP+ reductase. Electrons and hydrogen ions are added to NADP+ to form NADPH. This reducing agent is transported to the Calvin cycle to react with glycerate 3-phosphate, along with ATP to form glyceraldehyde 3-phosphate, the basic building block from which plants can make a variety of substances.

OXYGEN EVOLUTION

Oxygen evolution is the process of generating molecular oxygen through chemical reaction. Mechanisms of oxygen evolution include the photolysis of water during oxygenic photosynthesis, electrolysis of water into oxygen and hydrogen, and electrocatalytic oxygen evolution from oxides and oxoacids.

Oxygen Evolution in Nature

Photolytic oxygen evolution is the fundamental process by which breathable oxygen is generated in earth's biosphere. The reaction is part of the light-dependent reactions of photosynthesis in cyanobacteria and the chloroplasts of green algae and plants. It utilizes the energy of light to split a water molecule into its protons and electrons for photosynthesis. Free oxygen is generated as a waste product of this reaction, and is released into the atmosphere.

Biochemical Reaction

Photolytic oxygen evolution occurs via the light-dependent oxidation of water to molecular oxygen and can be written as the following simplified chemical reaction:

$2H_2O\ 4^{e-} + 4H^+ + O_2$

The reaction requires the energy of four photons. The electrons from the oxidized water molecules replace electrons in the P_{680} component of photosystem II which have been removed into an electron transport chain via light-dependent excitation and resonance energy transfer onto plastoquinone Photosytem II therefore has also been referred to as water-plastoquinone oxido-reductase. The protons are released into the thylakoid lumen, thus contributing to the generation of a proton gradient across the thylakoid membrane. This proton gradient is the driving force for ATP synthesis via photophosphorylation and coupling the absorption of light energy and photolysis of water to the creation of chemical energy during photosynthesis.

Oxygen-evolving Complex

Oxygen evolution by water oxidation during photosynthesis. The jagged lines represent four photons oxidizing the central cluster of the oxygen evolving complex by exciting and removing four electrons through a cycle of *S-states*.

Water oxidation is catalyzed by a manganese-containing enzyme complex associated with thylakoid membranes known as the oxygen evolving complex (OEC) or water-splitting complex. Manganese is an important cofactor, and calcium and chloride are also required for the reaction to occur.

X-ray crystallography studies have recently provided detailed models of the structure of the oxygen-evolving complex and its manganese cluster Based on structural and spectroscopic experiments, oxygen evolution involves a core three-plus-one cluster of three manganese ions and one calcium ion, with one additional manganese, which are oxidized via intermediate states called *S-states*. The O-O bond of molecular oxygen is formed between manganese-ligated oxygen atoms at the most oxidized, or S4, state.

Evolution of Oxygen Evolution

Oxygen production during photosynthesis evolved on earth around 2.7 to 3.5 billion years ago. Oxygen was not only a waste product of this reaction, but was also toxic to many metabolic processes such as nitrogen fixation. Consequently, it was released into the atmosphere as a means of detoxification. This contributed to the conversion of earth's atmosphere from anaerobic to its current aerobic composition, triggering the Oxygen Catastrophe and the evolution of aerobic metabolism utilizing the oxygen that was released by photosynthetic organisms as part of the oxygen cycle.

History of Discovery

It wasn't until the end of the 18th century that Joseph Priestley discovered by accident the ability of plants to "restore" air that had been "injured" by the burning of a candle. He followed up on the experiment by showing that air "restored" by vegetation was *"not at all inconvenient to a mouse."* He was later awarded a medal for his discoveries that: *". . . no vegetable grows in vain . . . but cleanses and purifies our atmosphere."* Priestley's experiments were followed up by Jan Ingenhousz, a Dutch physician, who showed that "restoration" of air only worked in the presence of light and green plant parts.

Ingenhousz suggested in 1796 that CO_2 (carbon dioxide) is split during photosynthesis to release oxygen, while the carbon combined with water to form carbohydrates. While this hypothesis was attractive and reasonable and thus widely accepted for a long time, it was later proven incorrect. Graduate student C.B. Van Niel at Stanford University found that purple sulfur bacteria reduce carbon to carbohydrates, but accumulate sulfur instead of releasing oxygen. He boldly proposed that in analogy to the sulfur bacteria forming elemental sulfur from H_2S (hydrogen sulfide), plants would form oxygen from H_2O (water). In 1937, this hypothesis was corroborated by the discovery that plants are capable of producing oxygen in the absence of CO_2. This discovery was made by Robin Hill, and subsequently the light-driven release of oxygen in the absence of CO_2 was called the *Hill reaction*. Our current knowledge of the mechanism of oxygen evolution during photosynthesis was further established in experiments tracing isotopes of oxygen from water to oxygen gas.

Technological Oxygen Evolution

Oxygen evolution occurs as a byproduct of hydrogen production via electrolysis of water. While oxygen production is not the main focus of industrial applications of water electrolysis, it becomes essential for life support systems in situations that require the generation of oxygen for air revitalization. Human exploration of regions that lack breathable oxygen, such as the deep sea or outer space, requires means of reliably generating oxygen apart from earth's atmosphere. Submarines and spacecraft utilize either an electrolytic mechanism (water or solid oxide electrolysis) or chemical oxygen generators as part of their life support equipment.

9

Photosensitization

INTRODUCTION

The novel anti-inflammatory drug benzydamine has been shown to photosensitize the reduction of Nitro Blue Tetrazolium, ferricytochrome-c and copper (II) bathocuproinedisulphonate in aqueous solutions (pH 7.4, 30°C) when irradiated with UV light at its maximum absorption wavelength of 308 nm. The reduction reactions all proceed most efficiently when the solutions are deoxygenated, clearly indicating that direct electron transfer occurs from the excited state of the sensitizer to the substrate. In aerated solutions the reduction reactions are slower and are partially inhibited by superoxide dismutase, suggesting that superoxide anion could be involved as an intermediate when oxygen is present. Benzydamine also photosensitizes the oxidation of l-histidine and 2, 5-dimethylfuran by the singlet oxygen pathway in aerated solutions. The ability of benzydamine to participate as sensitizer in several types of photochemical reaction is relevant to the observed clinical photosensitivity of the drug.

In addition to the direct responses to UVA and UVB exposure, the human system can be subjected to sunlight-caused effects similar to an exaggerated sunburn but mediated through exogenous photosensitizers such as prescription medication. In general, the generation of an adverse photosensitivity response can be postulated to involve one or more of the pathways shown in Fig. 9.1.

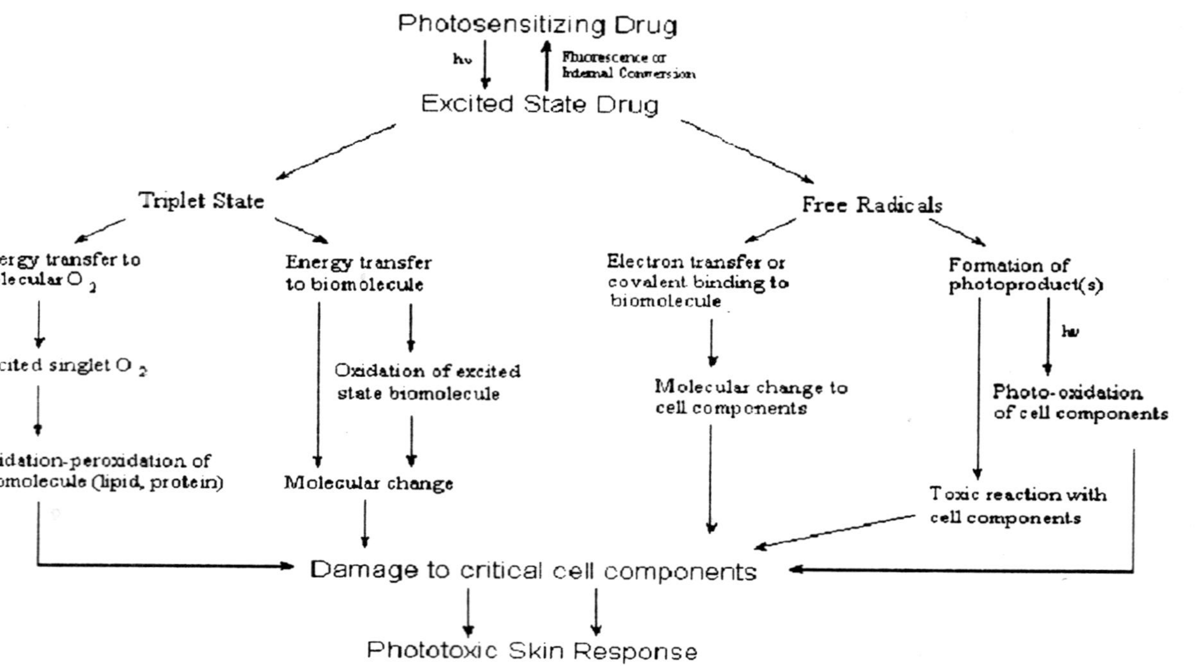
Photosensitizing Drug
hν
Fluorescence or Internal Conversion
Excited State Drug
Triplet State
Free Radicals
Energy transfer to molecular O_2
Energy transfer to biomolecule
Electron transfer or covalent binding to biomolecule
Formation of photoproduct(s)
Excited singlet O_2
Oxidation of excited state biomolecule
Molecular change to cell components
hν
Photo-oxidation of cell components
Oxidation-peroxidation of biomolecule (lipid, protein)
Molecular change
Toxic reaction with cell components
Damage to critical cell components
Phototoxic Skin Response

Fig. 9.1

Photosensitization by drugs was being investigated in our laboratory by studying the properties of their photoexcited states, and free radical and singlet oxygen mediated photooxidation reactions which ensue in model biological systems. The photooxidation of susceptible biological substrates by singlet oxygen or free radical pathways is widely believed to lead to the initiation of the adverse responses. However, it is possible that, in conditions of low oxygen concentration, or when the sensitizer is located close to susceptible substrates, direct electron transfer from sensitizer to substrate may become a significant contributor. Oxygen may also be involved as an electron carrier in the form of the superoxide anion radical.

Nitro Blue Tetrazolium (NBT) and (ferri)cytochrome-*c* (Cyt) are reagents which are frequently used to detect the occurrence of the superoxide anion radical in an enzymic or photosensitized reaction. Superoxide can be formed by electron transfer from a donor to molecular oxygen but it is quenched by NBT or Cyt which are thereby reduced to diformazan and ferrocytochrome-c, respectively. The detection of superoxide is confirmed when addition of the enzyme superoxide dismutase (SOD) causes a decrease in production of diformazan from NBT. The possibility exists for direct electron transfer from the photosensitizer to NBT if its rate is competitive with that of superoxide formation. This is more likely when oxygen is in short supply.

Another compound which can be used for this test is the copper (II) complex with a bifunctional ligand such as bathocuproine disulphonic acid (BCDS) which shows SOD-like activity . The electron transfer to the Cu(II)-BCDS complex resulting in reduction to the Cu(I) complex is detectable by a spectrophotometric change.

A number of photosensitizing drugs previously tested in this laboratory have been shown to undergo photoionization and/or are active in free radical generation. They may also participate in electron transfer processes. We have already reported that 6-mercaptopurine reacts with NBT or p-nitroso-dimethylaniline when irradiated in an oxygen-free solution The antibacterial drug sulfamethoxazole was also found to participate in photo-initiated electron transfer to NBT and Cyt . In this paper the use of NBT, Cyt and BCDS is described for the investigation of electron transfer mechanisms involving the drug benzydamine. This compound is a unique non-steroidal anti-inflammatory agent which also possesses local anaesthetic properties. However, many cases of photosensitivity reactions following topical application or oral ingestion of benzydamine have been reported . Its chemical structure is given with the UV absorption spectrum in Fig. 9.2.

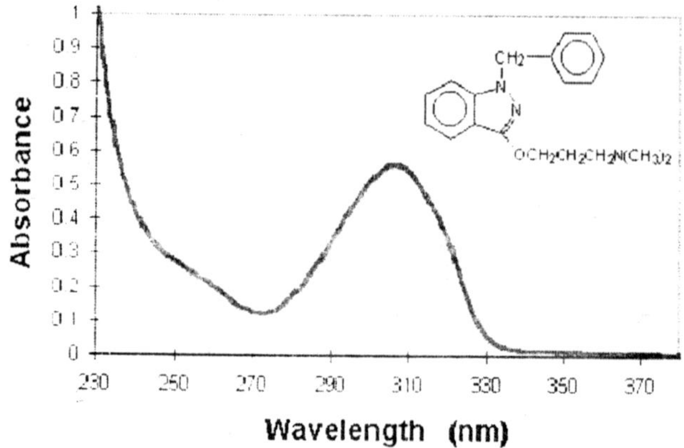

Fig. 9.2: Structure and absorption spectra of benzydamine (1 × 10-4 M) in aqueous buffer solutions.

Methods

Benzydamine hydrochloride was kindly supplied as the pure substance by 3M Pharmaceuticals Pty Ltd, Sydney, and used as received. Buffered aqueous solutions (phosphate, 0.05 M, pH 7.4) containing Cyt (0.1 mg/ml) NBT (5.0×10^{-5} M) or BCDS (0.1 mM) with 50 mM benzydamine were flushed with gas (N_2 or O_2 as required) for 40 minutes before irradiating. The samples were contained in stoppered cylindrical quartz vessels of 10 mm pathlength and irradiated with a 400 W medium pressure mercury arc (Applied Photophysics Ltd, UK) through a 2 mm Pyrex glass filter (Corning O-53). The photoreduction of NBT was followed as a function of the irradiation time by determining the increase in absorbance at 560 nm due to the diformazan product. The extent of reduction of Cyt was determined by measuring, as a function of irradiation time, the differences in absorbance between the maximum at 550 nm and the minimum near 535 nm. The reduction of Cu(II)-BCDS was monitored at 484 nm.

RESULTS

Reduction of Nitro Blue Tetrazolium

When NBT in N_2- or O_2-flushed solution was irradiated in the absence of a photosensitizer, no detectable reaction occurred. With the addition of benzydamine in a N_2-flushed solution the reduction of NBT could be seen by the appearance of the diformazan at 560 nm. Figure 9.3 shows the reaction photosensitized by benzydamine as a function of time. The reaction occurred to a lesser extent when an air-saturated solution was employed under otherwise identical conditions, and was completely inhibited when the solution was flushed with oxygen before irradiation. Thus it appears that the reduction of NBT photosensitized by benzydamine can be a direct reaction from the excited state of the sensitizer to NBT which molecular oxygen is able to inhibit.

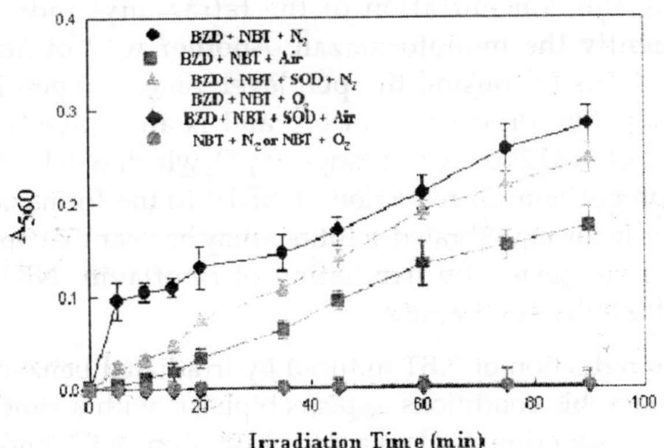

Fig. 9.3: Photoreduction of NBT (5.0 × 10-5 M) sensitized by benzydamine (5.0 × 10-5 M) in aqueous buffer solution pH 7.0 and 30°C

There are, however, a number of competing processes in which oxygen can be involved, such as singlet oxygen or superoxide formation. These were investigated as follows. Addition of SOD to the reaction mixture irradiated under air-saturated conditions completely inhibited the formation of diformazan, implying the involvement of superoxide in the reaction when some oxygen was present. The question arises as to why oxygen saturation inhibits the reaction completely when superoxide is diagnosed as being present under air-saturated conditions. It is suggested that the reaction is oxygen-concentration dependent. This behaviour is similar to that observed with 6-mercaptopurine but different from that with sulfamethoxazole or a-terthienyl However, all results are consistent with the report that direct elevation of pO_2 will suppress the photochemical reduction of NBT to formazan by simple mass action

The first step in the reduction of NBT^{2+} to the monoformazan MF^+ by O_2.- is the production in (1) of the tetrazoinyl radical which then disproportionates by reaction (2). The monoformazan then reacts to the end product diformazan by the same sequence of reactions. Addition of SOD shifts (1) to the left by decreasing the steady state concentration of $O^{2.-}$ thereby reducing the concentration of the tetrazoinyl radical and subsequently the monoformazan. Another way of affecting reaction (1) is by raising the pO_2 level. Since oxygen is both essential to the production of $O_2^{.-}$ and is an inhibitor of the reduction of $NBT2^+$ there is a range of pO_2 which will be optimal for the photochemical reduction of NBT^{2+} to the formazan. The pO_2 value in air-equilibrated solutions may be near that optimum value, as suggested by irradiation of riboflavin, NBT^{2+} and tetramethylethylenediamine.

The reduction of NBT induced by irradiated benzydamine under anaerobic conditions appears biphasic with a short rapid initial stage superimposed on a slower reaction. SOD diminished the extent of reduction only in this initial stage with little influence on the latter part. Similar results were also reported for photoreduction of NBT by riboflavin and methionine and 6-mercaptopurin . This result suggests that under anaerobic conditions, two mechanisms may be considered - either direct electron transfer from excited sensitizer to NBT, or initial expulsion of one electron from the excited sensitizer to the medium, and subsequent reaction of the solvated electron with the reagent. In both cases, the possibility that the reactions were actually due to one or more photoproducts could not be excluded, since the UV spectrum of benzydamine has changed after this period of irradiation time. All the above results implicate electron transfer from the excited states of benzydamine.

CYTOCHROME *c* REDUCTION

Ferricytochrome c is an electron acceptor which differs from the NBT reduction in that only one electron is required to reduce the fully oxidized "ferri" form to the fully reduced "ferro" form . From the redox potentials for O_2.- [E° (O_2/O_2.-) = -330 mv, E'° (O_2.-, H^+/H_2O_2) = 940 mv] and cytochrome *c* [E'' (Fe^{III} cytochrome *c* /Fe^{II} cytochrome *c*) = 260 mv] , it is evident that, on thermodynamic considerations alone, O_2.- could either oxidize or reduce ferricytochrome *c*. In the presence of $Na_2S_2O_4$ and O_2, $O_2^{.-}$ reduces ferricytochrome *c* in 100 % yield After capturing one electron, Cyt (III) will be reduced to Cyt (II) producing an absorption band at 550 nm. The absorption spectra of oxidized and reduced cytochrome *c* are very characteristic and distinct, and this spectral change provides a sensitive and convenient assay for an electron transfer reaction. It has been used to detect the superoxide anion in aerated condition and electron release from a sensitizer in deaerated conditions. Such electron transfer has been demonstrated with benzo[a]pyrene , a-terthienyl [anthracene, 2-chloro-3,11-tridecadiene-5,7,9-triyn-1-ol and sulfamethoxazole When oxygen- or air- or nitrogen- saturated samples containing benzydamine and cytochrome *c* were kept in the dark or when cytochrome *c* alone was irradiated, no spectral changes were observed. When the experiments were performed under continuous irradiation in a oxygen-saturated solution containing benzydamine 5.0×10^{-5} M and cytochrome *c* (0.1 mg/mL) in pH 7.4 phosphate buffer, there was an increase in absorbance at 550 nm. The difference in absorbance (A_{550} - A_{535}) plotted as a function of irradiation time in Fig. 9.4, shows a steady increase with time, reaching a plateau after about 10 min irradiation. However by adding sodium hydrosulfite to the

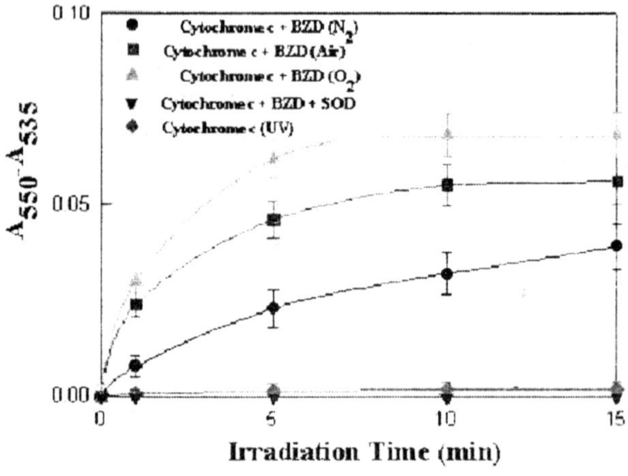

Fig. 9.4: Photoreduction of cytochrome *c* (0.1 mg/mL) sensitized by benzydamine (5.0×10^{-5} M). Each value shown is the average of three determinations

irradiated medium, a further increase in the absorbance of the 550 nm peak was observed, with conversion of all the Cyt (III) to Cyt (II). This means that cytochrome *c* was not the limiting reagent. There appear to be other factors inhibiting this reduction process. One possibility is that cytochrome *c* itself introduces an inner filter effect by light absorption at 365 nm, although this usually happens at cytochrome *c* concentrations higher than 0.1 mg/mL. Upon addition of SOD (300 units/mL) which competes with Cyt (III) for reaction with superoxide anion, the reduction was totally suppressed, implying that photoreduction of cytochrome *c* induced by benzydamine is a photodynamic process, in which the superoxide anion radical was involved in the photoinduced transfer of an electron to cytochrome *c*. This result is similar to that described for a-terthienyl in which the growth of the 550 nm band due to electron transfer induced by a-terthienyl to cytochrome *c* increased until it reached a constant value.

When the SOD experiment was carried out with benzydamine under air-saturated conditions, the extent of cytochrome *c* reduction decreased, but showed the same trend as observed under oxygen-saturated conditions. This suggests there may be intermediates involved in addition to superoxide anion. In fact, benzydamine can also participate in the photoreduction of cytochrome *c* by a mechanism which does not require oxygen. Under anaerobic conditions, cytochrome *c* photoreduction induced by benzydamine was demonstrated to take place at a rate faster than for cytochrome *c* alone, but slower than the photosensitized reduction under aerobic conditions, suggesting a direct reaction with cytochrome *c*. These results are similar to those reported with benzo[a]pyrene and anthracene in which the photoreduction was greater in the presence of oxygen than in its absence, and the photoreduction in the presence of oxygen was greatly diminished by addition of the enzyme SOD, indicating superoxide anion involvement.

The results obtained aerobically in the presence and absence of SOD prove that electronically excited benzydamine preferentially transfers electrons to cytochrome *c* using oxygen as an intermediate, even though the reduction can take place in its absence. This is probably because the dissolved oxygen concentration in air saturated water is 235 mM at 30°C representing a large molar excess over cytochrome *c* (8 mM) in the environment of each electronically excited molecule of benzydamine.

Reduction of Copper(II) Complex with SOD-like Activity

The knowledge that the enzyme Cu/Zn-SOD controls superoxide via disproportionation of O_2.- into O_2 and H_2O_2 suggests that some other compounds which are known as efficient catalysts of the dismutation process can be used to replace SOD to detect the involvement of electron transfer. For example, copper(II) is able to capture electrons in the presence of a complexing agent such as bathocuproinedisulphonic acid disodium salt hydrate (BCDS) which stabilizes the reduced copper as a copper(I) complex. The reaction course can be monitored by spectrophotometry at 484 nm which is the maximum absorption of the complex Cu(I)BCDS. The SOD-like activity of the copper system with BCDS has been verified by an indirect method using cytochrome *c* assay. When Cu(II)Cl2,

BCDS and benzydamine were mixed with cytochrome *c* under oxygen-saturated conditions, the reduction of cytochrome *c* was completely suppressed.

When benzydamine 1.0 × 10^{-4} M in phosphate buffered saline (PBS) solution was mixed with 0.1 mM BCDS and 30 mM copper sulphate, and irradiated both under aerobic and anaerobic conditions, an absorption change at 484 nm was observed. It was found that, under N_2, the reaction goes slightly faster than under O2. The absorption increased with the time of irradiation until all the copper (II) was transformed to the copper(I) complex (Fig. 9.5).

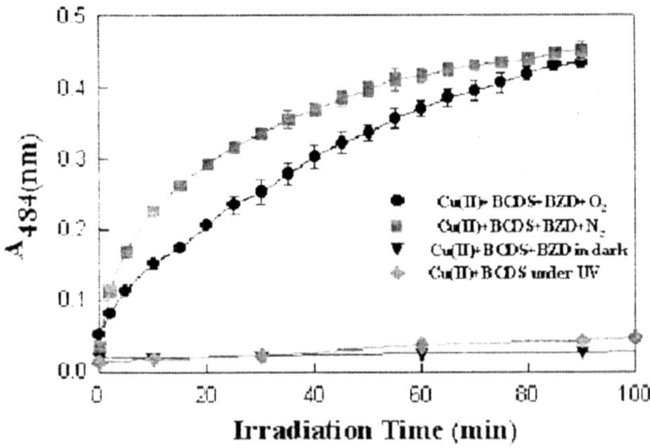

Fig. 9.5: Formation of copper(I) BCDS complex induced by benzydamine (1 × 10^{-4} M) as a function of irradiation time at pH 7.4 and 30°C

When Cu(II) and BCDS were irradiated without benzydamine or the solution of benzydamine and Cu(II) and BCDS was kept in the dark, no absorption at 484 nm was observed after 100 min. This result provided further confirmation that an electron transfer reaction induced by benzydamine had occurred. The slower rate in the presence compared to the absence of oxygen could be due to oxygen quenching the excited state of benzydamine. This is consistent with the results obtained through experiments with NBT and cytochrome *c*.

Singlet Oxygen Mediated Photooxidation

In separate experiments, benzydamine was found to photosensitize the oxidation of l-histidine or 2,5-dimethylfuran in air-saturated solutions. These compounds are substrates for singlet oxygen, and although they are not completely specific, a firm indication of the participation of singlet oxygen was found.

The photooxidation reaction was followed by measuring the depletion of oxygen with a Clarke-type oxygen electrode. Fig. 9.6 shows the effect of increasing concentration of benzydamine on the oxygen uptake rate of pH 7 buffered solution in the presence and absence of substrates when irradiated by the medium pressure mercury arc through a glass filter. Each result is the mean of triplicate measurements. In the presence of DF, or histidine, increased oxygen uptake was observed. The rate of oxygen consumption increased linearly at low benzydamine concentrations, then the rate gradually approached a plateau value, implying that all the radiation has been absorbed, and the light intensity becomes rate determining. No oxygen uptake was detected when DF or histidine was irradiated in the absence of drug. When benzydamine was added into the solution and kept out of the light, there was no detectable dark reaction. Upon irradiation, benzydamine was itself oxidized, the extent increasing with increasing concentration. At each concentration, the oxygen uptake rate was linear with time, *i.e.*, zero order, indicating that the absorption of light was the rate-limiting factor. The participation of singlet oxygen was also indicated by the fact that addition of 0.01 M azide ion, a $^1O_2$ quencher in aqueous solutions, reduced the photooxidation rate by 75%.

Variation of pH of the solution showed that the cation form of benzydamine (pKa 9.2) is 3-fold more efficient as a photosensitizer than the neutral form. Although the site of ionization in benzydamine is the alkylamino group which is distant from the absorbing chromophore, the difference in oxygen uptake capability between the base and its conjugate acid shows that the positive charge on the molecule is favourable for energy transfer to molecular oxygen. This is a similar pH dependence to that of the fluorescence yield of benzydamine and suggests that the free amino group quenches the excited state to some extent.

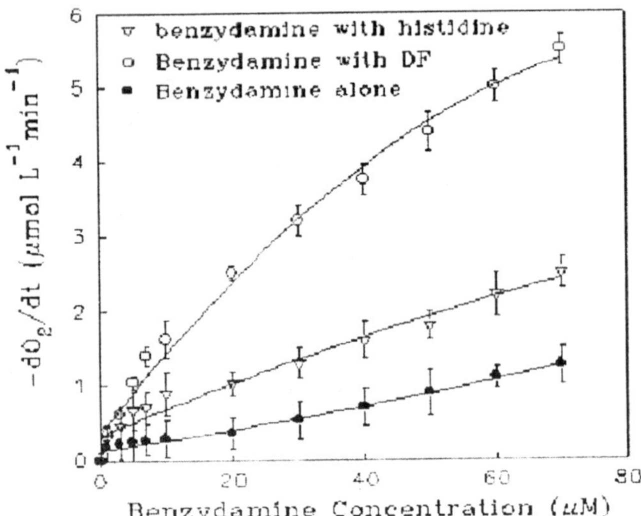

Fig. 9.6: Oxygen uptake rates as a function of benzydamine concentration at pH 7 and 30°C using 2,5-dimethylfuran (DF, 2 mM) and histidine (2 mM) as substrates

CONCLUSION

When benzydamine is irradiated in aerated solutions competing pathways of photosensitization are possible, namely, singlet oxygen mediated oxidation, and electron transfer mechanisms. The relative importance of each would be dependent of the presence of the relevant substrates near to the photoexcited benzydamine molecule. The electron transfer process is more likely to be an important factor in the photobiological activity of benzydamine in conditions of low oxygen concentrations.

Riboflavin is an important vitamin and well-known photosensitizer. However, under some conditions, its photosensitizing action is affected by the inhibiting action of photooxidation products. We observed this phenomenon in the course of oxidation of glycyl-tryptophan (Gly-Trp).

Peptide Gly-Trp is a convenient model for study of protein photooxidation. Photooxidation of Gly-Trp is accompanied with chemiluminescence (CL).

The study was carried out in a model system based on the measurement of photoinduced CL in aqueous solution. Chemiluminescence occurs due to decay of one of the oxidation products, dioxetane (D). Decay of the dioxetane results in product (P) in excited state:

$$D \rightarrow P, P^* \rightarrow P + hn_{CL}$$

This product is presumably, kynurenin-type compound.

D, L - kynurenin

Methods

Photochemiluminescence was measured with the specially designed set. The figure shows the scheme of measurement. The solution of Gly-Trp and riboflavin (phosphate buffer, pH 7.0) was poured into cuvette C_1 and was irradiated with high pressure mercury lamp under stirring. Wavelength of 436 nm was cut off with glass filters. Immediately after irradiation a portion of solution was pumped into cuvette C_2 of a high-sensitive photometric unit. Photocurrent of chemiluminescence was measured by recorder. The CL half-life was about 70 s at room temperature.

CL decay kinetic curves were measured in continuous irradiation mode with pumping of small portion of solution into photometric unit every minute and pumping back 10 s after.

When Rose Bengal and erythrosine were used as sensitizers, solutions were exposed to light of 546 nm.

KINETICS OF PHOTOSENSITIZED CL

The decrease in CL intensity during irradiation after the period of initial growth was found earlier in the case of direct irradiation of Gly-Trp solutions in UVA range without sensitizer. Further investigation showed the similar effect for RF-sensitized CL.

The effect is not due to sensitizer "fading" or Gly-Trp consumption: addition of fresh Gly-Trp or sensitizer portion slightly increases the intensity but does not restore initial CL level (Fig. 9.2). The effect is more pronounced at higher RF concentrations (Fig. 9.1) as well as at higher Gly-Trp concentrations. Addition of small quantity of preliminary irradiated solution to experimental solution before irradiation caused marked decrease in the initial intensity.

The data obtained allow us to conclude that the decrease in CL intensity is due to inhibiting action of product(s) formed in the course of irradiation. This assumption is in a good agreement with concentration effects: the more are Gly-Trp and RF concentrations or light intensity, the higher is a product formation rate; hence, the stronger is inhibition.

Inhibiting effect of products strongly depends on sensitizer nature. For example, no effect was observed when Rose Bengal (RB) or erythrosine were used as sensitizers , although the inhibiting products were formed. It was shown by addition of a small portion of Gly-Trp solution, preliminary irradiated with RB, to fresh RF-containing Gly-Trp solution before irradiation. The CL intensity in this case was lower than in control experiment. It is clear that the difference between RF and RB is due to the difference in mechanisms of occurring of CL.

There are different primary processes resulting in CL. Under direct excitation of tryptophan chromophore photo-ionization takes place . Peroxide radicals of Gly-Trp (RO2.) and superoxide are formed . It is assumed that reaction between peroxide radical and superoxide results in dioxetane (D):

$$RO_2\cdot + O_2^{-\cdot} \rightarrow D \; k_r \qquad (1)$$

The latter decays with excited kynurenin-type product *P formation:

$$D \rightarrow P, {}^*P \; k_d \qquad (2)$$

This product is responsible for chemiluminescence emission.

In the case of sensitized CL the light is absorbed by sensitizer molecule. When triplet excited states are generated there are two possible pathways for CL. The first includes singlet oxygen reaction with Gly-Trp that results in dioxetane formation. Another pathway includes reaction of sensitizer in triplet excited state with Gly-Trp. Then in the presence of oxygen peroxide radical of Gly-Trp (RO_2.) and superoxide are formed. The reaction of those results in dioxetane formation by reaction (1) as in the case of direct excitation. Of course, other ways, such as reactions of sensitizer in singlet excited state could not be excluded.

Riboflavin-sensitized CL

The CL pathway includes the following stages:

photo-initiation:

RF $\rightarrow$ ^{3}RF $\rightarrow$ (+Gly-Trp, +O_2) $\rightarrow$ RO_2., O_2-. w_i,

dioxetane formation by reaction (1) with reaction rate w_D, dioxetane decay by reaction (2).

Inhibiting effect of one (or a few) of oxidation products (In) is realized by reaction

$$\text{In} + O_2^{-\cdot} \rightarrow \ldots k_{In} \qquad (3)$$

and may be also by reaction with RO_2.:

$$\text{In} + RO_2^{-\cdot} \rightarrow \ldots 'k_{In} \qquad (4)$$

This retards dioxetane formation as concentration of inhibiting product increases. Based on equation:

$$d[\text{D}]/dt = w_D - k_d[\text{D}]$$

when the influence of products is not accounted dioxetane concentration as well as CL intensity increases with time as

$$I \sim [\text{D}] \sim 1 - \exp(-k_d t).$$

If the rate of product formation is proportional to dioxetane concentration

$$d[\text{P}]/dt \sim [\text{D}]$$

then concentration of product depends on time as

$$[\text{P}] \sim k_d t + \exp(-k_d t) - 1.$$

This dependance supposes initial delay and asimptote [P] ~ t at long times. The "tails" of kinetic curves from Fig. 1 fit to lines when plotted in coordinates I_{max}/I versus time It suggests that concentration of product-inhibitor linearly rises with time.

It is known that the main products of tryptophan oxidation are kynurenin derivatives. We studied the effect of DL-kynurenin on CL of Gly-Trp (with RF as sensitizer)

Concentration of half-inhibition was equal to 5.10^{-6} M, rate constant for reaction of DL-kynurenin with superoxide was assessed as 1.10^{-7} $M^{-1}s^{-1}$

The oxidation rate in the experiments was too low to detect kynurenin by spectrophotometry. However, at higher irradiation doses we measured increase in absorbanse at 364 nm, corresponding kynurenin formation

In the course of oxidation of peptide glycyltryptophan, photosensitized by riboflavin, some oxidation products, presumably kynurenin derivatives, act as inhibitors. The inhibiting effect is due to reaction of inhibitor with superoxide and/or peroxide radicals of peptide. Singlet oxygen oxidation is not affected, although inhibiting products are formed. The effect depends on riboflavin and peptide concentrations and exposure dose.

10

Sunlight

INTRODUCTION

Sunlight, in the broad sense, is the total spectrum of the electromagnetic radiation given off by the Sun. On Earth, sunlight is filtered through the atmosphere, and the solar radiation is obvious as daylight when the Sun is above the horizon. This is usually during the hours known as *day*. Near the poles in summer, sunlight also occurs during the hours known as *night* and in the winter at the poles sunlight may not occur at any time. When the direct radiation is not blocked by clouds, it is experienced as sunshine, a combination of bright light and heat. Radiant heat directly produced by the radiation of the sun is different from the increase in atmospheric temperature due to the radiative heating of the atmosphere by the sun's radiation. Sunlight may be recorded using a sunshine recorder, pyranometer or pyrheliometer. The World Meteorological Organization defines sunshine as direct irradiance from the Sun measured on the ground of at least 1120 $W{\cdot}m^{-2}$.

Direct sunlight has a luminous efficacy of about 93 lumens per watt of radiant flux, which includes infrared, visible, and ultra-violet light. Bright sunlight provides luminance of approximately 100,000 candela per square meter at the Earth's surface.

Sunlight is a Key Factor in the Process of Photosynthesis

To calculate the amount of sunlight reaching the ground, both the elliptical orbit of the earth and the earth's atmosphere have to be taken into account. The extraterrestrial solar illuminance (E_{ext}), corrected for the elliptical orbit by using the day number of the year, known as the Julian date (Jd), is: $E_{ext}=E_{sc}$ $(1 + 0.034 * cos(2pi(Jd - 2)/365)$

The solar illuminance constant (E_{sc}), is equal to 128 Klux. The direct normal illuminance, (E_{dn}), corrected for the attenuating effects of the atmosphere is given by: $E_{dn}=E_{ext}*e^{-cm}$

Where c is the atmospheric extinction coefficient and m is the relative optical air mass.

SOLAR CONSTANT

Solar irradiance spectrum at top of atmosphere, on a linear scale and plotted against wavenumber.

The solar constant is the amount of incoming solar electromagnetic radiation per unit area, measured on the outer surface of Earth's atmosphere in a plane perpendicular to the rays. The solar constant includes all types of solar radiation, not just the visible light. It is measured by satellite to be roughly 1366 watts per square meter (W/m^2), though this fluctuates by about 6.9% during a year (from 1412 W/m^2 in early January to 1321 W/m^2 in early July) due to the earth's varying distance from the Sun, and typically by much less than one part per thousand from day to day. Thus, for the whole Earth (which has a cross section of 127,400,000 km^2), the power is 1.740×10^{17} W, plus or minus 3.5%. The solar constant does not remain constant over long periods of time (see Solar variation). The approximate average value cited 1366 W/m^2, is equivalent to 1.96 calories per minute per square centimeter, or 1.96 langleys (Ly) per minute.

The Earth receives a total amount of radiation determined by its cross section ($p \cdot R_E^2$), but as it rotates this energy is distributed across the entire surface area ($4 \cdot p \cdot R_E^2$). Hence the average incoming solar radiation (sometimes called the solar irradiance), taking into account the angle at which the rays strike and that at any one moment half the planet does not receive any solar radiation, is one-fourth the solar constant (approximately 342 W/m^2). At any given moment, the amount of Solar radiation received at a location on the Earth's surface depends on the state of the atmosphere and the location's latitude.

The solar constant includes all wavelengths of solar electromagnetic radiation, not just the visible light. It is linked to the apparent magnitude of the Sun, -26.8, in that the solar constant and the magnitude of the Sun are two methods of describing the apparent brightness of the Sun, though the magnitude only measures the visual output of the Sun.

In 1884, Samuel Pierpont Langley attempted to estimate the Solar constant from Mount Whitney in California. By taking readings at different times of day, he attempted to remove effects due to atmospheric absorption. However, the value he obtained, 2903 W/m^2, was still too great. Between 1902 and 1957, measurements by Charles Greeley Abbot and others at various high-altitude sites found values between 1322 and 1465 W/m^2. Abbott proved that one of Langley's corrections was erroneously applied. His results varied between 1.89 and 2.22 calories (1318 to 1548 W/m^2), a variation that appeared to be due to the Sun and not the Earth's atmosphere.

The angular diameter of the Earth as seen from the Sun is approximately 1/11,000 radians, meaning the solid angle of the Earth as seen from the sun is approximately 1/140,000,000 steradians. Thus the Sun emits about two billion times the amount of radiation that is caught by Earth, in other words about 3.86 × 1026 watts

The actual brightness of sunlight that would be observed at the surface depends also on the presence and composition of an atmosphere. For example Venus' thick atmosphere reflects more than 60% of the solar light it receives. The actual illumination of the surface is about 5000-10000 lux, comparable to that of Earth during a dark, very cloudy day.

Sunlight on Mars would be more or less like daylight on Earth wearing sunglasses, and as can be seen in the pictures taken by the rovers, there is enough diffuse sky radiation that shadows would not seem particularly dark. Thus it would give perceptions and "feel" very much like Earth daylight.

For comparison purposes, sunlight on Saturn is somewhat slightly brighter than Earth sunlight on the average sunset or sunrise. Even on Pluto the Sun would be still bright enough to almost match the average living room. To see the Sun shine as dim as the full Moon on the Earth, a distance of about 500 AU (~69 light-hours) is needed: there is only a handful of objects in the solar system known to orbit farther than such a distance, among them 90377 Sedna and (87269) 2000 OO67.

COMPOSITION

The spectrum of the Sun's solar radiation is close to that of a black body with a temperature of about 5,800 K. About half that lies in the visible short-wave part of the electromagnetic spectrum and the other half mostly in the near-infrared part. Some also lies in the ultraviolet part of the spectrum. When ultraviolet radiation is not absorbed by the atmosphere or other protective coating, it can cause a change in human skin pigmentation.

The spectrum of electromagnetic radiation striking the Earth's atmosphere is 100 to 10^6 nanometer (nm). This can be divided into five regions in increasing order of wavelengths:

- *Ultraviolet C or (UVC)* range, which spans a range of 100 to 280 nm. The term *ultraviolet* refers to the fact that the radiation is at higher frequency than violet light (and, hence also invisible to the human eye). Owing to absorption by the atmosphere very little reaches the Earth's surface (Lithosphere). This spectrum of radiation has germicidal properties, and is used in germicidal lamps.
- *Ultraviolet B or (UVB)* range spans 280 to 315 nm. It is also greatly absorbed by the atmosphere, and along with UVC is responsible for the photochemical reaction leading to the production of the Ozone layer.
- *Ultraviolet A or (UVA)* spans 315 to 400 nm. It has been traditionally held as less damaging to the DNA, and hence used in tanning and PUVA therapy for psoriasis.

- *Visible range or light* spans 400 to 700 nm. As the name suggests, it is this range that is visible to the naked eye.
- *Infrared range* that spans 700 nm to 106 nm [1 millimeter (mm)]. It is largely responsible for the warmth or heat that the sunlight carries. It is also divided into three types on the basis of wavelength:
    - Infrared-A: 700 nm to 1400 nm.
    - Infrared-B: 1400 nm to 3000 nm.
    - Infrared-C: 3000 nm to 1 mm.

CLIMATE EFFECTS

On Earth, solar radiation is obvious as daylight when the sun is above the horizon. This is during daytime, and also in summer near the poles at night, but not at all in winter near the poles. When the direct radiation is not blocked by clouds, it is experienced as *sunshine,* combining the perception of bright white light (sunlight in the strict sense) and warming. The warming on the body and surfaces of other objects is distinguished from the increase in air temperature.

The amount of radiation intercepted by a planetary body varies inversely with the square of the distance between the star and the planet. The Earth's orbit and obliquity change with time (over thousands of years), sometimes forming a nearly perfect circle, and at other times stretching out to an orbital eccentricity of 5% (currently 1.67%). The total insolation remains almost constant but the seasonal and latitudinal distribution and intensity of solar radiation received at the Earth's surface also varies. For example, at latitudes of 65 degrees the change in solar energy in summer & winter can vary by more than 25% as a result of the Earth's orbital variation. Because changes in winter and summer tend to offset, the change in the annual average insolation at any given location is near zero, but the redistribution of energy between summer and winter does strongly affect the intensity of seasonal cycles. Such changes associated with the redistribution of solar energy are considered a likely cause for the coming and going of recent ice ages.

LIFE ON EARTH

The existence of nearly all life on Earth is fueled by light from the sun. Most autotrophs, such as plants, use the energy of sunlight to turn air into simple sugars—a process known as photosynthesis. These sugars are then used as building blocks and in other synthetic pathways which allow the organism to grow.

Heterotrophs, such as animals, use light from the sun indirectly by consuming the products of autotrophs, either directly or by consuming other heterotrophs. The sugars and other molecular components produced by the autotrophs are then broken down, releasing stored solar energy, and giving the heterotroph the energy required for survival. This process is known as respiration.

In prehistory, humans began to further extend this process by putting plant and animal materials to other uses. They used animal skins for warmth, for example, or wooden weapons to hunt. These skills allowed humans to harvest more of the sunlight than was possible through glycolysis alone, and human population began to grow.

During the Neolithic Revolution, the domestication of plants and animals further increased human access to solar energy. Fields devoted to crops were enriched by inedible plant matter, providing sugars and nutrients for future harvests. Animals which had previously only provided humans with meat and tools once they were killed were now used for labour throughout their lives, fueled by grasses inedible to humans.

The more recent discoveries of coal, petroleum and natural gas are modern extensions of this trend. These fossil fuels are the remnants of ancient plant and animal matter, formed using energy from sunlight and then trapped within the earth for millions of years. Because the stored energy in these fossil fuels has accumulated over many millions of years, they have allowed modern humans to massively increase the production and consumption of primary energy. As the amount of fossil fuel is large but finite, this cannot continue indefinitely, and various theories exist as to what will follow this stage of human civilization (e.g. alternative fuels, Malthusian catastrophe, new urbanism, peak oil).

Cultural Aspects

Many people find direct sunlight to be too bright for comfort, especially when reading from white paper upon which the sun is directly shining. Indeed, looking directly at the sun can cause permanent vision damage. To compensate for the brightness of sunlight, many people wear sunglasses. Cars, many helmets and caps are equipped with visors to block the sun from direct vision when the sun is at a low angle.

Prism Splitting Light

In colder countries many people prefer sunnier days and often avoid the shade. In hotter countries the converse is true; during the midday hours many people prefer to stay inside to remain cool. If they do go outside, they seek shade which may be provided by trees, parasols, and so on.

Sunshine is often blocked from entering buildings through the use of walls, window blinds, awnings, shutters or curtains.

SUNBATHING

Sunbathing is a popular leisure activity in which a person sits or lies in direct sunshine. People often sunbathe in comfortable places where there is ample sunlight. Some common places for sunbathing include beaches, open air swimming pools, parks, gardens, and sidewalk cafés. Sunbathers typically wear limited amounts of clothing or some simply go nude. An alternative some use to sunbathing is to use a sunbed that generates ultraviolet light and can be used indoors regardless of outdoor weather conditions and amount of sun light.

For many people with pale or brownish skin, one purpose for sunbathing is to darken one's skin color (get a sun tan) as this is considered in some cultures to be beautiful, associated with outdoor activity, vacations or holidays, and health. Some people prefer nude sunbathing so that an "all-over" or "even" tan can be obtained.

Skin tanning is achieved by an increase in the dark pigment inside skin cells called melanocytes and it is actually an automatic response mechanism of the body to sufficient exposure to ultraviolet radiation from the sun or from artificial sunlamps. Thus, the tan gradually disappears with time, when one is no longer exposed to these sources.

Effects on Health

The body produces vitamin D from sunlight (specifically from the UVB band of ultraviolet light), and excessive seclusion from the sun can lead to deficiency unless adequate amounts are obtained through diet.

Excessive sunlight exposure has been linked to all types of skin cancer caused by the ultraviolet part of radiation from sunlight or sunlamps. Sunburn can have mild to severe inflammation effects on skin; this can be avoided by using a proper sunscreen cream or lotion or by gradually building up melanocytes with increasing exposure. Another detrimental effect of UV exposure is accelerated skin aging (also called skin photodamage), which produces a difficult to treat cosmetic effect. Some people are concerned that ozone depletion is increasing the incidence of such health hazards. A 10% decrease in ozone could cause a 25% increase in skin cancer.

A lack of sunlight, on the other hand, is considered one of the primary causes of seasonal affective disorder (SAD), a serious form of the "winter blues". SAD occurrence is more prevalent in locations further from the tropics, and most of the treatments (other than prescription drugs) involve replicating sunlight via sunlamps tuned to specific (visible, not ultra-violet) wavelengths of light or full-spectrum bulbs.

A recent study indicates that more exposure to sunshine early in a person's life relates to less risk from multiple sclerosis (MS) later in life

EARTH'S ATMOSPHERE

The Earth's atmosphere is a layer of gases surrounding the planet Earth that is retained by the Earth's gravity. Dry air contains roughly (by molar content – equivalent to volume, for gases) 78.08% nitrogen, 20.95% oxygen, 0.93% argon, 0.038% carbon dioxide, and trace amounts of other gases; but air also contains a variable amount of water vapor, on average around 1%. This mixture of gases is commonly known as air. The atmosphere protects life on Earth by absorbing ultraviolet solar radiation, warming the surface through heat retention (greenhouse effect), and reducing temperature extremes between day and night.

There is no definite boundary between the atmosphere and outer space. It slowly becomes thinner and fades into space. Three quarters of the atmosphere's mass is within 11 km of the planetary surface. An altitude of 120 km (~75 miles or 400,000 ft) marks the boundary where atmospheric effects become noticeable during re-entry. The Kármán line, at 100 km (62 miles or 328,000 ft), is also frequently regarded as the boundary between atmosphere and outer space.

Temperature and Layers

The temperature of the Earth's atmosphere varies with altitude; the mathematical relationship between temperature and altitude varies among five different atmospheric layers (ordered highest to lowest, the ionosphere is part of the thermosphere):

- **Exosphere**: from 500 – 1000 km (300 – 600 mi) up to 10,000 km (6,000 mi), free-moving particles that may migrate into and out of the magnetosphere or the solar wind exobase boundary.
- **Ionosphere**: the part of the atmosphere that is ionized by solar radiation. It plays an important part in atmospheric electricity and forms the inner edge of the magnetosphere. It has practical importance because, among other functions, it influences radio propagation to distant places on the Earth. It is located in the thermosphere and is responsible for auroras thermopause boundary.
- **Thermosphere**: from 80-85 km (265,000-285,000 ft) to 640+ km (400+ mi), temperature increasing with height mesopause boundary.

- **Mesosphere**: The mesosphere extends from about 50 km (160,000 ft) to the range of 80 to 85 km (265,000-285,000 ft), temperature decreasing with height. This is also where most meteors burn up when entering the atmosphere stratopause boundary.
- **Stratosphere**: From the Latin word "*stratus*" meaning a spreading out. The stratosphere extends from the troposphere's 7 to 17 km (23,000 – 60,000 ft) range to about 50 km (160,000 ft). Temperature increases with height. The stratosphere contains the ozone layer, the part of the Earth's atmosphere which contains relatively high concentrations of ozone. "Relatively high" means a few parts per million—much higher than the concentrations in the lower atmosphere but still small compared to the main components of the atmosphere. It is mainly located in the lower portion of the stratosphere from approximately 15 to 35 km (50,000 – 115,000 ft) above Earth's surface, though the thickness varies seasonally and geographically tropopause boundary.
- **Troposphere**: The troposphere is the lowest layer of the atmosphere; it begins at the surface and extends to between 7 km (23,000 ft) at the poles and 17 km (60,000 ft) at the equator, with some variation due to weather factors. The troposphere has a great deal of vertical mixing because of solar heating at the surface. This heating warms air masses, which makes them less dense so they rise. When an air mass rises, the pressure upon it decreases so it expands, doing work against the opposing pressure of the surrounding air. To do work is to expend energy, so the temperature of the air mass decreases. As the temperature decreases, water vapor in the air mass may condense or solidify, releasing latent heat that further uplifts the air mass. This process determines the maximum rate of decline of temperature with height, called the adiabatic lapse rate. The troposphere contains roughly 80% of the total mass of the atmosphere. Fifty percent of the total mass of the atmosphere is located in the lower 5.6 km of the troposphere.

The average temperature of the atmosphere at the surface of Earth is 15 °C (59 °F).

PRESSURE AND THICKNESS

The average atmospheric pressure, at sea level, is about 101.3 kilopascals (about 14.7 psi); total atmospheric mass is 5.1480×1018 kg.

Atmospheric pressure is a direct result of the total weight of the air above the point at which the pressure is measured. This means that air pressure varies with location and time, because the amount (and weight) of air above the earth varies with location and time. However the *average* mass of the air above a square meter of the earth's surface is known to the same high accuracy as the total air mass of 5148.0 teratonnes and area of the earth of 51007.2 megahectares, namely 5148.0/510.072 = 10.093 metric tonnes per square meter or 14.356 lbs (mass) per square inch. This is about 2.5% below the officially standardized unit atmosphere (1 atm) of 101.325 kPa or 14.696 psi, and corresponds to the mean pressure not at sea level but at the mean base of the atmosphere as contoured by the earth's terrain.

Were atmospheric density to remain constant with height the atmosphere would terminate abruptly at 7.81 km (25,600 ft.). Instead it decreases with height, dropping by 50% at an altitude of about 5.6 km (18,000 ft). For comparison: the highest mountain, Mount Everest, is higher, at 8.8 km, which is why it is so difficult to climb without supplemental oxygen. This pressure drop is approximately exponential, so that pressure decreases by approximately half every 5.6 km (whence about 50% of the total atmospheric mass is within the lowest 5.6 km) and by 63.2 % ($1 - 1/e = 1 - 0.368 = 0.632$) every 7.64 km, the average scale height of Earth's atmosphere below 70 km. However, because of changes in temperature, average molecular weight, and gravity throughout the atmospheric column, the dependence of atmospheric pressure on altitude is modeled by separate equations for each of the layers listed above.

Even in the exosphere, the atmosphere is still present (as can be seen for example by the effects of atmospheric drag on satellites).

The equations of pressure by altitude in the above references can be used directly to estimate atmospheric thickness. However, the following published data are given for reference:

- 50% of the atmosphere by mass is below an altitude of 5.6 km.
- 90% of the atmosphere by mass is below an altitude of 16 km. The common altitude of commercial airliners is about 10 km.
- 99.99997% of the atmosphere by mass is below 100 km. The highest X-15 plane flight in 1963 reached an altitude of 354,300 ft (108.0 km).

Therefore, most of the atmosphere (99.9997%) is below 100 km, although in the rarefied region above this there are auroras and other atmospheric effects.

Composition

Filtered air includes at least trace amounts of ten (or more) of the chemical elements. Substantial amounts of argon, nitrogen, and oxygen are present as elementary gases, as well as hydrogen (and additional oxygen) in water vapor (H_2O). Much smaller or trace amounts of elementary helium, hydrogen, iodine, krypton, neon, and xenon are also present, as well as carbon in carbon dioxide (CO_2), methane (CH_4), and carbon monoxide (CO). Many additional elements from natural sources may be present in tiny amounts in an unfiltered air sample, including contributions from dust, pollen and spores, sea spray, vulcanism, and meteoroids. Various industrial pollutants are also now present in the air, such as chlorine (elementary or in compounds), fluorine (in compounds), elementary mercury, and sulfur (in compounds such as sulfur dioxide [SO_2]).

The composition figures above are by volume-fraction (V%), which for ideal gases is equal to mole-fraction (that is, the fraction of total molecules). Although the atmosphere is not an ideal gas, nonetheless the atmosphere behaves enough like an ideal gas that the volume-fraction is the same as the mole-fraction for the precision given.

By contrast, *mass-fraction* abundances of gases will differ from the volume values. The mean molar mass of air is 28.97 g/mol, while the molar mass of helium is 4.00, and krypton is 83.80. Thus helium is 5.2 ppm by *volume-fraction*, but 0.72 ppm by *mass-fraction* ([4/29] × 5.2 = 0.72), and krypton is 1.1 ppm by *volume-fraction*, but 3.2 ppm by *mass-fraction* ([84/29] × 1.1 = 3.2).

Heterosphere

Below the turbopause at an altitude of about 100 km (not far from the mesopause), the Earth's atmosphere has a more-or-less uniform composition (apart from water vapor) as described above; this constitutes the homosphere. However, above about 100 km, the Earth's atmosphere begins to have a composition which varies with altitude. This is essentially because, in the absence of mixing, the density of a gas falls off exponentially with increasing altitude but at a rate which depends on the molar mass. Thus higher mass constituents, such as oxygen and nitrogen, fall off more quickly than lighter constituents such as helium, molecular hydrogen, and atomic hydrogen. Thus there is a layer, called the heterosphere, in which the earth's atmosphere has varying composition. As the altitude increases, the atmosphere is dominated successively by helium, molecular hydrogen, and atomic hydrogen. The precise altitude of the heterosphere and the layers it contains varies significantly with temperature.

In pre-history, the Sun's radiation caused a loss of the hydrogen, helium and other hydrogen-containing gases from early Earth, and Earth was devoid of an atmosphere. The first atmosphere was formed by outgassing of gases trapped in the interior of the early Earth, which still goes on today in volcanoes.

The density of air at sea level is about 1.2 kg/m^3(1.2 g/L). Natural variations of the barometric pressure occur at any one altitude as a consequence of weather. This variation is relatively small for inhabited altitudes but much more pronounced in the outer atmosphere and space because of variable solar radiation.

The atmospheric density decreases as the altitude increases. This variation can be approximately modeled using the barometric formula. More sophisticated models are used by meteorologists and space agencies to predict weather and orbital decay of satellites.

The average mass of the atmosphere is about 5 quadrillion metric tons or 1/1,200,000 the mass of Earth. According to the National Center for Atmospheric Research, "The total mean mass of the atmosphere is 5.1480×10^{18} kg. with an annual range due to water vapor of 1.2 or 1.5×10^{15} kg. depending on whether surface pressure or water vapor data are used; somewhat smaller than the previous estimate. The mean mass of water vapor is estimated as 1.27×10^{16} kg. and the dry air mass as 5.1352 $\pm 0.0003 \times 10^{18}$ kg."

Opacity

Rough plot of Earth's atmospheric transmittance (or opacity) to various wavelengths of electromagnetic radiation, including visible light.

The atmosphere has "windows" of low opacity, allowing the transmission of electromagnetic radiation. The optical window runs from around 300 nanometers (ultraviolet-C) at the short end up into the range the eye can use, the visible spectrum at roughly 400-700 nm, and continues up through the visual infrared to around 1100 nm, which is thermal infrared. There are also infrared and radio windows that transmit some infrared and radio waves. The radio window runs from about one centimeter to about eleven-meter waves.

Evolution of Earth's Atmosphere

The history of the Earth's atmosphere prior to one billion years ago is poorly understood and an active area of scientific research. The following discussion presents a plausible scenario.

The modern atmosphere is sometimes referred to as Earth's "third atmosphere", in order to distinguish the current chemical composition from two notably different previous compositions. The original atmosphere was primarily helium and hydrogen. Heat from the still-molten crust, and the sun, plus a probably enhanced solar wind, dissipated this atmosphere.

About 4.4 billion years ago, the surface had cooled enough to form a crust, still heavily populated with volcanoes which released steam, carbon dioxide, and ammonia. This led to the early "second atmosphere", which was primarily carbon dioxide and water vapor, with some nitrogen but virtually no oxygen. This second atmosphere had approximately 100 times as much gas as the current atmosphere, but as it cooled much of the carbon dioxide was dissolved in the seas and precipitated out as carbonates. The later "second atmosphere" contained largely nitrogen and carbon dioxide. However, simulations run at the University of Waterloo and University of Colorado in 2005 suggest that it may have had up to 40% hydrogen. It is generally believed that the greenhouse effect, caused by high levels of carbon dioxide and methane, kept the Earth from freezing.

One of the earliest types of bacteria was the cyanobacteria. Fossil evidence indicates that bacteria shaped like these existed approximately 3.3 billion years ago and were the first oxygen-producing evolving phototropic organisms. They were responsible for the initial conversion of the earth's atmosphere from an anoxic state to an oxic state (that is, from a state without oxygen to a state with oxygen) during the period 2.7 to 2.2 billion years ago. Being the first to carry out oxygenic photosynthesis, they were able to produce oxygen while sequestering carbon dioxide in organic molecules, playing a major role in oxygenating the atmosphere.

Photosynthesising plants later evolved and continued releasing oxygen and sequestering carbon dioxide. Over time, excess carbon became locked in fossil fuels, sedimentary rocks (notably limestone), and animal shells. As oxygen was released, it reacted with ammonia to release nitrogen; in addition, bacteria would also convert ammonia into nitrogen. But most of the nitrogen currently present in the atmosphere results from sunlight-powered photolysis of ammonia released steadily over the aeons from volcanoes.

As more plants appeared, the levels of oxygen increased significantly, while carbon dioxide levels dropped. At first the oxygen combined with various elements (such as iron), but eventually oxygen accumulated in the atmosphere, contributing to Cambrian explosion and further evolution. With the appearance of an ozone layer (ozone is an allotrope of oxygen) lifeforms were better protected from ultraviolet radiation. This oxygen-nitrogen atmosphere is the "third atmosphere". Between 200 and 250 million years ago, up to 35% of the atmosphere was oxygen (as found in bubbles of ancient atmosphere preserved in amber).

This modern atmosphere has a composition which is enforced by oceanic blue-green algae as well as geological processes. O_2 does not remain naturally free in an atmosphere but tends to be consumed (by inorganic chemical reactions, and by animals, bacteria, and even land plants at night), and CO_2 tends to be produced by respiration and decomposition and oxidation of organic matter. Oxygen would vanish within a few million years by chemical reactions, and CO_2 dissolves easily in water and would be gone in millennia if not replaced. Both are maintained by biological productivity and geological forces seemingly working hand-in-hand to maintain reasonably steady levels over millions of years.

Currently, anthropogenic greenhouse gases are increasing in the atmosphere and this is a causative factor in global warming.

AIR POLLUTION

Before desulfurization filters were installed, the emissions from this power plant in New Mexico contained excessive amounts of sulfur dioxide.

Air pollution is the human introduction into the atmosphere of chemicals, particulate matter, or biological materials that cause harm or discomfort to humans or other living organisms, or damages the environment Stratospheric ozone depletion is believed to be caused by air pollution (chiefly from chlorofluorocarbons)

Worldwide air pollution is responsible for large numbers of deaths and cases of respiratory disease. Enforced air quality standards, like the Clean Air Act in the United States, have reduced the presence of some pollutants. While major stationary sources are often identified with air pollution, the greatest source of emissions is actually mobile sources, principally the automobile Gases such as carbon dioxide, methane, and fluorocarbons contribute to global warming, and these gases, or excess amounts of some emitted from fossil fuel burning, have recently been identified by the United States and many other countries as pollutants.

Kyoto Protocol

Participation in the Kyoto Protocol, where dark green indicates countries that have signed and ratified the treaty, yellow is signed, but not yet ratified, grey is not yet decided and red is no intention of ratifying.

As of April 2008, 178 States have signed and ratified the Kyoto Protocol to the United Nations Framework Convention on Climate Change, aimed at combating global warming.

The Kyoto Protocol is a protocol to the international Framework Convention on Climate Change with the objective of reducing greenhouse gases in an effort to prevent anthropogenic climate change.

It was adopted for use on 11 December 1997 by the 3rd Conference of the Parties, which was meeting in Kyoto, and it entered into force on 16 February 2005. As of May 2008, 182 parties have ratified the protocol Of these, 36 developed C.G. countries (plus the EU as a party in its own right) are required to reduce greenhouse gas emissions to the levels specified for each of them in the treaty (representing over 61.6% of emissions from Annex I countries) with three more countries intending to participate One hundred thirty-seven (137) developing countries have ratified the protocol, including Brazil, China and India, but have no obligation beyond monitoring and reporting emissions. The United States is the only developed and industrialized western country that has not ratified the treaty but it is one of the significant greenhouse gas emitters.

BIOSPHERE

The biosphere is the broadest level of ecological study, the global sum of all ecosystems. From the broadest biophysiological point of view, the biosphere is the global ecological system integrating all living beings and their relationships, including their interaction with the elements of the lithosphere, hydrosphere, and atmosphere. This biosphere is postulated to have evolved, beginning through a process of biogenesis or biopoesis, at least some 3.5 billion years ago.

Origin and Use of the Term

While this concept has a geological origin, it is an indication of the impact of both Darwin and Maury on the earth sciences. The biosphere's ecological context comes from the 1920s (*see* Vladimir I. Vernadsky), preceding the 1935 introduction of the term "ecosystem" by Sir Arthur Tansley (see ecology history). Vernadsky defined ecology as the science of the biosphere. It is an interdisciplinary concept for integrating astronomy, geophysics, meteorology, biogeography, evolution, geology, geochemistry, hydrology and, generally speaking, all life and earth sciences.

Narrow Definition

Some life scientists and earth scientists use *biosphere* in a more limited sense. For example, geochemists define the biosphere as being the total sum of living organisms (the "biomass" or "biota" as referred to by biologists and ecologists). In this sense, the biosphere is but one of four separate components of the geochemical model, the other three being *lithosphere, hydrosphere,* and *atmosphere*. The narrow meaning used by geochemists is one of the consequences of specialization in modern science. Some might prefer the word *ecosphere,* coined in the 1960s, as all encompassing of both biological and physical components of the planet.

The Second International Conference on Closed Life Systems defined *biospherics* as the science and technology of analogs and models of Earth's biosphere; i.e., artificial Earth-like biospheres. Others may include the creation of artificial non-Earth biospheres — for example, human-centered biospheres or a native Martian biosphere — in the field of biospherics.

The concept that the biosphere is itself a living organism, either actually or metaphorically, is known as the Gaia hypothesis.

James Lovelock, an atmospheric scientist from the United Kingdom, proposed the Gaia hypothesis to explain how biotic and abiotic factors interact in the biosphere. This hypothesis considers Earth itself a kind of living organism. Its atmosphere, geosphere, and hydrosphere are cooperating systems that yield a biosphere full of life. in the early 1970s, Lynn Margulis, a microbiologist from the United States, added to the hypothesis specifically noting the ties between the biosphere and other Earth systems. For example, when carbon dioxide levels increase in the atmosphere, plants grow more quickly. As their growth continue, they remove more and more carbon dioxide from the atmosphere.

Many scientists are now devoting their careers to organizing new fields of study, such as geobiology and geomicrobiology, to examine these intriguing relationships.

Extent of Earth's Biosphere

Nearly every part of the planet, from the polar ice caps to the Equator, supports life of some kind. Recent advances in microbiology have demonstrated that microbes live deep beneath the Earth's terrestrial surface, and that the total mass of microbial life in so-called "uninhabitable zones" may, in biomass, exceed all animal and plant life on the surface. The actual thickness of the biosphere on earth is difficult to measure. Birds typically fly at altitudes of 650 to 2000 meters, and fish that live deep underwater can be found down to -8,372 meters in the Puerto Rico Trench.

There are more extreme examples for life on the planet: Rüppell's Vulture has been found at altitudes of 11,300 meters; Bar-headed Geese migrate at altitudes of at least 8,300 meters (over Mount Everest); Yaks live at elevations between 3,200 to 5,400 meters above sea level; mountain goats live up to 3,050 meters. Herbivorous animals at these elevations depend on lichens, grasses, and herbs but the biggest tree is the Tine palm or mountain coconut found 3,400 meters above sea level.

Microscopic organisms live at such extremes that, taking them into consideration puts the thickness of the biosphere much greater. Culturable microbes have been found in the Earth's upper atmosphere as high as 41km (Wainwright et al, 2003, in FEMS Microbiology Letters). It is unlikely, however, that microbes are active at such altitudes, where temperatures and air pressure are extremely low and ultraviolet radiation very high. More likely these microbes were brought into the upper atmosphere by winds or possibly volcanic eruptions. Barophilic marine microbes have been found at more than 10km depth in the Marianas Trench. Microbes are not limited to the air, water or the Earth's surface. Culturable thermophilic microbes have been extracted from cores drilled more than 5km into the Earth's crust in Sweden, from rocks between 65-75C. Temperature increases rapidly with increasing depth into the Earth's crust. The speed at which the temperature increases depends on many factors, including type of crust (continental vs. oceanic), rock type, geographic location, etc. The upper known limit of microbial is 122C (*Methanopyrus kandleri* Strain 116), and it is likely that the limit of life in the "deep biosphere" is defined by temperature rather than absolute depth.

Our biosphere is divided into a number of biomes, inhabited by broadly similar flora and fauna. On land, biomes are separated primarily by latitude. Terrestrial biomes lying within the Arctic and Antarctic Circles are relatively barren of plant and animal life, while most of the more populous biomes lie near the equator. Terrestrial organisms in temperate and Arctic biomes have relatively small amounts of total biomass, smaller energy budgets, and display prominent adaptations to cold, including world-spanning migrations, social adaptations, homeothermy, estivation and multiple layers of insulation.

Specific Biospheres

When the word *Biosphere* is followed by a number, it is usually referring to a specific system or number. Thus:

- Biosphere 1 - The planet Earth
- Biosphere 2 - A laboratory in Arizona which contains 3.15 acres (13,000 m^2) of closed ecosystem.
- BIOS-3 was a closed ecosystem at the Institute of Biophysics in Krasnoyarsk, Siberia, in what was then the Soviet Union.

- Biosphere J (CEEF, Closed Ecology Experiment Facilities) - An experiment in Japan.

BIOME

A biome is a climatically and geographically defined area of ecologically similar climatic conditions such as communities of plants, animals, and soil organisms, and are often referred to as ecosystems. Biomes are defined based on factors such as plant structures (such as trees, shrubs, and grasses), leaf types (such as broadleaf and needleleaf), plant spacing (forest, woodland, savanna), and climate. Unlike ecozones, biomes are not defined by genetic, taxonomic, or historical similarities. Biomes are often identified with particular patterns of ecological succession and climax vegetation.

The biodiversity characteristic of each biome, especially the diversity of fauna and subdominant plant forms, is a function of abiotic factors and the biomass productivity of the dominant vegetation. In terrestrial biomes, species diversity tends to correlate positively with net primary productivity, moisture availability, and temperature

Ecoregions are grouped into both biomes and ecozones. A fundamental classification of biomes is into:

- Terrestrial (land) biomes;
- Freshwater biomes; and
- Marine biomes.

Biomes are often given local names. For example, a Temperate grassland or shrubland biome is known commonly as *steppe* in central Asia, *prairie* in North America, and *pampas* in South America. Tropical grasslands are known as *savanna* in Australia as well as Southern Africa where in Afrikaans it is known as *veldt*. Sometimes an entire biome may be targeted for protection, especially under an individual nation's Biodiversity Action Plan.

Climate is a major factor determining the distribution of terrestrial biomes. Among the important climatic factors are:

- *Latitude:* Arctic, boreal, temperate, subtropical, tropical.
- *Humidity:* humid, semi-humid, semi-arid, and arid.
- *Seasonal variation:* Rainfall may be distributed evenly throughout the year or be marked by seasonal variations.
- *Dry summer, wet winter:* Most regions of the earth receive most of their rainfall during the summer months; Mediterranean climate regions receive their rainfall during the winter months.
- *Elevation:* Increasing elevation causes a distribution of habitat types similar to that of increasing latitude.

Biodiversity generally increases away from the poles towards the equator and increases with humidity. The most widely used systems of classifying biomes correspond to latitude (or temperature zoning) and humidity. More info needed on the estuBailey system.

Robert G. Bailey developed a biogeographical classification system for the United States in a map published in 1975. Bailey subsequently expanded the system to include the rest of North America in 1981 and the world in 1989. The Bailey system is based on climate and is divided into four domains (Polar, Humid Temperate, Dry, and Humid Tropical), with further divisions based on other climate characteristics (subarctic, warm temperate, hot temperate, and subtropical; marine and continental; lowland and mountain).

- **100 Polar Domain**
  - 120 Tundra Division
  - M120 Tundra Division-Mountain Provinces
  - 130 Subarctic Division
  - M130 Subarctic Division-Mountain Provinces
- **200 Humid Temperate Domain**
  - 210 Warm Continental Division
  - M210 Warm Continental Division-Mountain Provinces

- 220 Hot Continental Division
- M220 Hot Continental Division-Mountain Provinces
- 230 Subtropical Division
- M230 Subtropical Division-Mountain Provinces
- 240 Marine Division
- M240 Marine Division-Mountain Provinces
- 250 Prairie Division
- 260 Mediterranean Division
- M260 Mediterranean Division-Mountain Provinces

- **300 Dry Domain**
  - 310 Tropical/Subtropical Steppe Division
  - M310 Tropical/Subtropical Steppe Division-Mountain Provinces

World Wide Fund for Nature System

A team of biologists convened by the World Wide Fund for Nature (WWF) developed an ecological land classification system that identified fourteen biomes, called **major habitat types,** and further divided the world's land area into 825 terrestrial ecoregions. This classification is used to define the Global 200 list of ecoregions identified by the WWF as priorities for conservation. The WWF major habitat types are as follows:

- Tundra (Arctic)
- Boreal forests/taiga (subarctic, humid)
- Temperate coniferous forests (temperate, humid to semi-humid)
- Temperate broadleaf and mixed forests (temperate, humid)
- Temperate grasslands, savannas, and shrublands (temperate, semi-arid)
- Mediterranean forests, woodlands, and shrub (temperate warm, semi-humid to semi-arid with winter rainfall)

- Tropical and subtropical coniferous forests (tropical and subtropical, semi-humid)
- Tropical and subtropical moist broadleaf forests (tropical and subtropical, humid)
- Tropical and subtropical dry broadleaf forests (tropical and subtropical, semi-humid)
- Tropical and subtropical grasslands, savannas, and shrublands (tropical and subtropical, semi-arid)
- Montane grasslands and shrublands (alpine or montane climate)
- Deserts and xeric shrublands (temperate to tropical, arid)
- Mangrove (subtropical and tropical, salt water inundated)
- Flooded grasslands and savannas (temperate to tropical, fresh or brackish water inundated)

Freshwater Biomes

According to the World Wildlife Fund, the following are classified as freshwater biomes:

- Large lakes
- Large river deltas
- Polar freshwaters
- Montane freshwaters
- Temperate coastal rivers
- Temperate floodplain rivers and wetlands
- Temperate upland rivers
- Tropical and subtropical coastal rivers
- Tropical and subtropical floodplain rivers and wetlands
- Tropical and subtropical upland rivers
- Xeric freshwaters and endorheic basins
- Oceanic islands

Marine Biomes

Global 200 marine major habitat types

- Polar
- Temperate shelves and sea
- Temperate upwelling
- Tropical upwelling
- Tropical coral

Other marine habitat types

- Continental shelf
- Littoral/Intertidal zone
- Coral reef
- Kelp forest
- Pack ice
- Hydrothermal vents
- Cold seeps
- Benthic zone
- Pelagic zone
- Neritic zone

Anthropogenic Biomes

Humans have fundamentally altered global patterns of biodiversity and ecosystem processes. As a result, vegetation forms predicted by conventional biome systems are rarely observed across most of Earth's land surface. Anthropogenic biomes provide an alternative view of the terrestrial biosphere based on global patterns of sustained direct human interaction with ecosystems, including agriculture, human settlements, urbanization, forestry and other uses of land. Anthropogenic biomes offer a new way forward in ecology and conservation by recognizing the irreversible coupling of human and ecological systems at global scales and moving us toward an understanding how best to live in and manage our biosphere and the anthropogenic biosphere we live in.

Major Anthropogenic Biomes

- Dense Settlements
- Villages
- Croplands
- Rangelands
- Forested

Other Biomes

The Endolithic biome, consisting entirely of microscopic life in rock pores and cracks, kilometers beneath the surface, has only recently been discovered and does not fit well into most classification schemes.

11

Photomorphogenesis

INTRODUCTION

Light has profound effects on the development of plants. The light-mediated changes in plant growth and development are called photomorphogenesis. The most striking effects of light are observed when a germinating seedling emerges from the soil and is exposed to light for the first time.

Normally the seedling radicle (root) emerges first from the seed, and the shoot appears as the root becomes established. Later, with growth of the shoot (particularly when it merges into the light) there is increased secondary root formation and branching. This coordinated progression of developmental responses are early manifestations of correlative growth phenomena where the root affects the growth of the shoot and vice versa. To a large degree, these coordinated differential growth responses are hormone mediated.

In the absence of light, plants develop an etiolated growth pattern. Etiolation of the seedling adapts it to emerging from the soil.

Comparison of dark-grown (etiolated) and light-grown (de-etiolated) seedlings

Etiolated characteristics	De-etiolated characteristics
Distinct "apical hook" (dicot) or coleoptile (monocot)	Apical hook opens or coleoptile splits open

No leaf growth	Leaf growth promoted
No chlorophyll	Chlorophyll produced
Rapid stem elongation	Stem elongation suppressed
Limited radial expansion of stem	Radial expansion of stem
Limited root elongation	Root elongation promoted
Limited production of lateral roots	Lateral rootdevelopment accelerated

The developmental changes characteristic of photomorphogenesis shown by de-etiolated seedlings, are induced by light. Typically, plants are responsive to wavelengths of light in the blue, red and far-red regions of the spectrum through the action of several different photosensory systems. The photoreceptors for red and far-red wavelengths are know as phytochromes. There are at least 5 members of the phytochrome family of photoreceptors. There are several blue light photoreceptors.

Photoreceptor Systems in Plants Red/far-red Systems: Phytochrome

- Plants use phytochrome to detect and respond to red and far-red wavelengths.
- Phytochromes are proteins with a light absorbing pigment attached (chromophore).
- The chromophore is a linear tetrapyrrole called phytochromobilin.

The phytochrome apoprotein is synthesized in the Pr form. Upon binding the chromophore, the holoprotein becomes sensitive to light. If it absorbs red light it will change conformation to the biologically active Pfr form. The Pfr form can absorb red light and switch back to the Pr form.

Most plants have multiple phytochromes encoded by different genes. The different forms of phytochrome control different responses but there is also a lot of redundancy so that in the absence of one phytochrome, another may take on the missing functions.

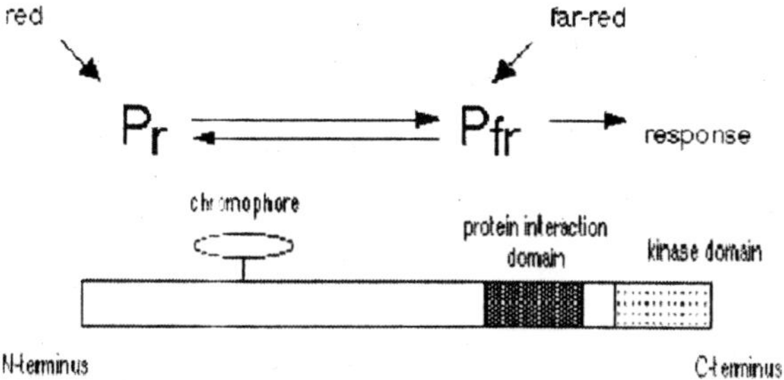

Schematic Diagram of Phytochrome protein

Arabidopsis has 5 phytochromes - PHYA, PHYB, PHYC, PHYD, PHYE

Molecular analyses of phytochrome and phytochrome-like genes in higher plants, ferns, mosses, algae, and photosynthetic bacteria have shown that phytochromes evolved from prokaryotic photoreceptors that predated the origin of plants.

BLUE LIGHT SYSTEMS

As for the red/far-red system, plants contain multiple blue light photoreceptors which have different functions.

Based on studies with action spectra, mutants and molecular analyses, it has been determined that higher plants contain at least 4, and probably 5, different blue light photoreceptors.

Cryptochromes were the first blue light receptors to be isolated and characterized from any organism. The proteins use a flavin as a chromophore. The cryptochromes have evolved from microbial DNA-photolyase, an enzyme that carries out light-dependent repair of UV damaged DNA.

Two cryptochromes have been identified in plants:

- Cryptochromes control stem elongation, leaf expansion, circadian rhythms and flowering time.
- In addition to blue light, cryptochromes also perceive long wavelength UV irradiation (UV-A).

Phototropin is the blue light photoreceptor that controls phototropism. It also uses flavin as chromophore. Only one phototropin has been identified so far (NPH1). Phototropin also perceives long wavelength UV irradiation (UV-A) in addition to blue light.

Experiments indicate that a 4th blue light receptor exists that uses a carotenoid as a chromophore. This new photoreceptor controls blue light induction of stomatal opening. However, the gene and protein have not yet been found.

Other blue light responses exist that seem to function in plants that are missing the cryptochrome, phototropin and carotenoid photoreceptors suggesting that at least one more will be found.

Since the cryptochromes were discovered in plants, several labs have identified homologous genes and photoreceptors in a number of other organisms, including humans, mice and flys. It appears that in mammals and flys, the cryptochromes function in entrainment of the biological clock. Indeed, in flys, a cryptochrome may be a functional part of the clock mechanism.

UV SYSTEMS

Based on various responses to UV light, it is assumed that there are UV-specific photoreceptors.

Plants can sense light direction, quality (wavelength), intensity and periodicity. Light induces phototropism, photomorphogenesis, chloroplast differentiation and various other responses such as flowering and germination.

Light quality is mainly sensed by the presence of different light receptors specific for different wavelengths. The red/far red photoreceptors are called phytochrome. There are at least 2 classes of blue light receptors; cryptochrome recognizes blue, green and UV-A light, while phototropin perceives blue light.

ANALYSIS OF PHOTOMORPHOGENESIS

Plants exhibit different growth habits in dark and light. In the dark they have elongated stems, undifferentiated chloroplasts and unexpanded leaves. This is called skotomorphogenesis. Photomorphogenesis (light grown) involves the inhibition of stem elongation, the differentiation of chloroplasts and accumulation of chlorophyll, and the expansion of leaves. Thus the same stimulus causes opposite effects on cell elongation in leaves and stems. Photomorphogenesis can be induced by red, far red and blue light.

Much of our knowledge of light perception and signaling has come from genetic analyses of photomorphogenesis. Essentially two types of mutant screens have been performed:

Screens for mutants that look dark grown even in the light (ie. insensitive or unresponsive to light). These are often designated as hy mutants for hypocotyl elongated, a dark grown character. These mutants have identified the known light receptors and a couple other genes that function as positive regulators of the light responses.

Screens for mutants that look light grown even in the dark. These are designated as cop for constitutive photomorphogenic or det for de-etiolated (etiolated is a term used to describe the dark grown habit). These recessive mutants are epistatic to hy mutants indicating that they function as negative regulators of signal transduction steps downstream of the receptors. In other words, because loss of function mutations allow photomorphogenic development in the absence of the inducing signal (light), the normal function of the DET and COP genes is to repress photomorphogenesis in the dark.

PHYTOCHROME

Phytochrome is a protein containing a covalently attached chromophore. Phytochrome exists in 2 interconvertable conformations with different absorbtion spectra. Pfr absorbs far red and is generally the biologically active conformation. Pr absorbs red. Absorbtion of red light converts Pr to Pfr while absorbtion of far red converts Pfr to Pr. Phytochrome responses are classically defined by their red/far red reversibility. For example, lettuce seeds require light to germinate. Red light induces germination but if followed by a pulse of far red light, germination is inhibited. It also contains a domain resembling a protein kinase and has been shown to autophosphorylate, however the functional significance of this in light signal transduction is unknown.

Phytochrome can measure light quality because if light contains more red than far red light (as is the case in daytime sunlight), most phytochrome will be in the Pfr form. Phytochrome mediates a variety of photomorphogenic phenomena including leaf expansion and inhibition of stem elongation. One classic example is in the shade avoidance response of shade intolerant plants. Foliage readily absorbs red light and so in the shade of another plant there is higher amounts of far red light which will drive phytochrome to the Pr form. Pr does not inhibit stem elongation which allows shaded plants to elongate and grow to reach the sunlight.

Arabidopsis contains 5 phytochrome genes, *PHYA-E,* each with distinct but often overlapping functions. PHYA is photolabile while PHYB is light stable. Different phytochromes control different plant processes in response to different intensities of light. Responses are classified as hi-irradiance (HI), low fluence (LF) or very low fluence (VLF) based on the intensity of light required to trigger the response. Phytochrome studies are further complicated by the fact that different species show different sets of responses to given light conditions. Therefore it is difficult to draw generalizations about the functions of different phytochromes.

In the dark, phytochrome is localized to the cytoplasm. In the light, phytochrome translocates to the nucleus. Phytochrome physically interacts with at least one transcription factor, PIF3, in a light dependent manner. PIF3 is required for phytochrome mediated photomorphogenesis because antisense lines show elongated hypocotyls in the light (ie. decreased light response). PIF3 binds to G-BOX elements of light regulated genes and is required for phytochrome mediated regulation of several genes.

Cryptochrome

Blue, green and UVA light are all perceived by a receptor called cryptochrome. It is a flavin protein with 2 chromophores attached, one for green, one for blue. There are 2 cryptochrome genes in arabidopsis, *CRY1* and 2. Again, they have distinct but overlapping functions. hy4/cry1 is a nonphotomorphogenic mutant defective for the blue light receptor. CRY proteins appear constitutively nuclear, although there are indications that there may be some CRY functions in the cytoplasm too.

Cryptochrome action requires the presence of phytochrome because some phytochrome mutants are non-photomorphogenic in blue or green light. However the cryptochrome mutant is photomorphogenic in red light (with far red reversibility) indicating that phytochrome action does not require cryptochrome. Evidence suggests that phytochrome and cryptochrome physically interact. CRY protein can be phosphorylated in vitro by the protein kinase activity of PHY-A. Furthermore, PHYB and CRY2 interact in plant extracts and exhibit FRET in plant cells (Mas et al., 2000). CRY1 and 2 also appear to directly interact with COP1, a factor involved in the negative regulation of photomorphogenesis in the dark (see below) (Yang et al., 2001).

HY5

HY5 is a transcription factor that is a key regulator of photomorphogenesis in both the phy and cry pathways. *hy5* mutants show a dark grown habit in the light indicating that *HY5* is required for light response. In the light, the level of HY5 protein increases and in the dark, it declines. *HY5* mRNA increases in response to phytochrome activation but the gene does not have a G-BOX suggesting it may not be regulated by PIF3.

COP1 and the COP9 Signalsome

A large number of mutants were identified with *constitutive photomorphogenesis (cop)* or *de-etiolated (det)* phenotypes. These mutants look light grown in the dark and therefore are thought to function as negative regulators of light signal transduction pathways (i.e. the signal transduction pathways are active in the absence of light). Many of these mutants encode proteins that form a large complex called the COP9 signalsome (CNS). The CNS is a nuclear complex that is similar to the 26S proteosome. This is a proteolytic complex that degrades ubiquitinated proteins.

COP1 is another protein that inhibits photomorphogenesis. In the dark, COP1 is present in the nucleus, but in the light it is only found in the cytoplasm. COP1 contains a ring-finger, which is a feature of many E3 ubiquitin ligases, which are involved in targeting proteins for 26S proteosome-mediated degredation. COP1 interacts with HY5, and in the dark HY5 protein levels show a decline that is dependent on COP1 and CNS. Thus it is hypothesized that in the dark, COP1 targets HY5 for degredation by CNS.

As mentioned, CRY1 and 2 are constitutively nuclear and also interact with COP1. It is hypothesized that CRY binding inhibits COP1 activity. Another factor that functions in phytochrome signal transduction, SPA1, also interacts with COP1. Thus, all photomorphogenetic signal transduction pathways appear to converge on COP1. Inhibition of COP1 may then allow accumulation of HY5 and the response to light.

HORMONES

Several plant hormones are thought to be involved in photomorphogenesis. Pfr appears to inhibit the sensitivity of hypocotyls to GA, thus in the dark when Pfr is depleted, the hypocotyls become more sensitive to GA and elongate. However brassinolide has a much more central role in photomorphogenesis. Several of the cop and det mutants are rescuable with exogenous brassinolide (ie. they show dark grown habit in the dark). The det2 mutant of arabidopsis was cloned and has sequence homology with mammalian steroid 5a-reductase. This suggests a function of the DET2 gene product in a particular step of brassinolide biosynthesis. Similarly, the CPD gene shows a cop-like mutant phenotype, was cloned and has sequence similarity to another enzyme in testosterone biosynthesis, suggesting it functions in another step of the brassinolide pathway. Rescue experiments with pathway intermediates were consistent with the proposed functions of these genes. Thus, light appears to control photomorphogenesis by downregulating brassinolide production.

Phototropin

Another blue light receptor is called phototropin. Arabidopsis contains 2 phototropins. These are involved in phototropism, but not photomorphogensis (hypocotyl elongation).

Phototropism is the directional growth in response to directional light. Shoots are positively phototropic (grow toward light) while roots often show negative phototropism (grow away from light). Directional light causes a redistribution of auxin in shoot tissues such that the side away from the light accumulates higher levels and grows faster, causing bending toward the light. Phototropism is controlled mainly by blue/UVA light. However the receptor is different from cryptochrome because the cry mutants are still phototropic. A mutant identified as nonphototropic hypocotyl1 (nph1), acts at the level of light perception in the phototropic response but still retains normal photomorphogenetic responses. This gene was recently renamed *PHOTOTROPIN1* (*PHOT1*). Arabidopsis contains two phototropin genes, *PHOT1* and *PHOT2*. PHOT protein is another flavoprotein containing 2 flavin mononucleotide chromophores. The protein also contains a protein kinase domain and blue light induces PHOT kinase activity.

12

Photoreceptor Cell

INTRODUCTION

A photoreceptor, or photoreceptor cell, is a specialized type of neuron (nerve cell) found in the eye's retina that is capable of phototransduction. The great biological importance of photoreceptors is that as cells they convert light (electromagnetic radiation) into the beginning of a chain of biological processes. More specifically, the photoreceptor absorbs photons from the field of view, and through a specific and complex biochemical pathway, signals this information through a change in its membrane potential.

For hundreds of years, photoreceptors in vertebrates were thought to be of only two main classes. The two classic photoreceptors are rods and cones, each contributing information used by the visual system to form a representation of the visual world, sight.

The novel third photoreceptor is a recently discovered class of photosensitive ganglion cells. These cells, found in the inner retina, have dendrites and long axons projecting to the pretectum (midbrain), the suprachiasmatic nucleas in the hypothalamus, and the lateral geniculate (thalamus).

There are major functional differences between the rods and cones. Cones are adapted to detect colors, and function well in bright light; rods are more sensitive, but do not detect colour well, being adapted for low light. In humans there are three different types of cone - responding respectively to short (blue), medium (green) and long (yellow-red) light. The human retina contains about 120 million rod cells and 6 million cone cells. The number and ratio of rods to cones varies among species, dependent on whether an animal is primarily diurnal or nocturnal. Certain owls have a tremendous number of rods in their retinas — the eyes of the tawny owl are approximately

Described here are vertebrate photoreceptors. Invertebrate photoreceptors in organisms such as insects and molluscs are different in both their morphological organization and their underlying biochemical pathways.

ANATOMY OF A ROD CELL

Rod and cone photoreceptors have the same basic structure. Closest to the visual field (and farthest from the brain) is the axon terminal, which releases a neurotransmitter called glutamate to bipolar cells. Farther back is the cell body, which contains the cell's organelles. Farther back still is the inner segment, a specialized part of the cell full of mitochondria. The chief function of the inner segment is to provide ATP (energy) for the sodium-potassium pump. Finally, closest to the brain (and farthest from the field of view) is the outer segment, the part of the photoreceptor that absorbs light. Outer segments are actually modified cilia that contain disks filled with opsin, the molecule that absorbs photons, as well as voltage-gated sodium channels.

The membranous photoreceptor protein *opsin* contains a pigment molecule called *retinal.* In rod cells, these together are called rhodopsin. In cone cells there are different types of opsins that combine with retinal to form pigments called photopsins. Three different classes of photopsins in the cones react to different ranges of light frequency, a differentation which eventually allows the visual system to distinguish color. The function of the photoreceptor cell is to convert the light energy of the photon into a form of energy communicable to the nervous system and readily usable to the organism: this conversion is called signal transduction.

The opsin found in the photosensitive ganglion cells of the retina that are involved in various reflexive responses of the brain and body to the presence of (day) light, such as the regulation of circadian rhythms, pupillary reflex and other non-visual responses to light, is called melanopsin. Atypical in vertebrates, melanopsin functionally resembles invertebrate opsins. In structure, it is an opsin, a retinylidene protein variety of G-protein-coupled receptor.

When light activates the melanopsin signaling system, the melanopsin-containing ganglion cells discharge nerve impulses which are conducted through their axons to specific brain targets. These targets include the olivary pretectal nucleus (a center responsible for controlling the pupil of the eye) and, through the retinohypothalamic tract (RHT), the suprachiasmatic nucleus of the hypothalamus (the master pacemaker of circadian rhythms). Melanopsin ganglion cells are thought to influence these targets by releasing from their axon terminals the neurotransmitters glutamate and pituitary adenylate cyclase activating polypeptide (PACAP).

Humans

In humans, the visual system uses millions of photoreceptors to view, perceive, and analyze the visual world. With the exception of melanopsin-containing photosensitive ganglion cells, ocular photoreceptors are the only neurons in humans capable of phototransduction. All photoreceptors in humans are found either in the outer nuclear layer in the retina at the back of each eye, while the bipolar and ganglion cells that transmit information from photoreceptors to the brain are in front of them. This inverted arrangement significantly reduces acuity, as light must travel through the axons and cell bodies of other neurons before reaching the photoreceptors. The retina contains two specializations to deal with this issue. First, a region at the center of the retina, called the fovea, containing only photoreceptors, is used for high visual acuity. Second, each retina contains a blind spot, an area where axons from the ganglion cells can go back through the retina to the brain.

For hundreds of years humans were thought to have only two types of photoreceptors: rods and cones. Both transduce light into a change in membrane potential through the same signal transduction pathway. However, they differ in the nature of the opsin they contain, and their function. Rods are used primarily to see at low levels of light, while cones are used to determine color, depth, and intensity. Furthermore, there are three types of cones, which differ in the spectrum of wavelengths of photons over which they absorb. Because cones respond to both the wavelength and intensity of light, a single cone cannot tell color; instead, color vision requires interactions of more than one type of cone, primarily by comparing responses across different cone types.

It is important to note that older textbooks will not contain yet the most exciting discovery in the field of human photoreception and vision. In 2007 a novel ganglion cell photoreceptor class was isolated in rodless coneless humans, where it was found to mediate circadian rhythms, behaviour, pupil reactions, and most exciting of all a component to unconscious and conscious sight. These non-rod non-cone photoreceptors in general are discussed later below.

PHOTOTRANSDUCTION

Phototransduction is the complex process whereby the energy of a photon is used to change the inherent membrane potential of the photoreceptor — and thereby signal to the nervous system that light is in the visual field.

Dark Current

Unstimulated (in the dark), the voltage-gated sodium channels in the outer segment are open because cyclic GMP (cGMP) is bound to them. Hence, positively charged sodium ions enter the photoreceptor, depolarizing it to about -40 mV (resting potential in other nerve cells is usually -65 mV). This depolarizing current is often known as dark current.

SIGNAL TRANSDUCTION PATHWAY

The signal transduction pathway is the mechanism by which the energy of a photon signals a mechanism in the cell that leads to its electrical polarization. This polarization ultimately leads to either the transmittance or inhibition of a neural signal that will be fed to the brain via the optic nerve. The steps, or signal transduction pathway, in the vertebrate eye's rod and cone photoreceptors are then:

— The rhodopsin or iodopsin in the outer segment absorbs a photon, changing the configuration of a retinal Schiff base cofactor inside the protein from the cis-form to the trans-form, causing the retinal to change shape.

— This results in a series of unstable intermediates, the last of which binds stronger to the G protein in the membrane and activates transducin, a protein inside the cell. This is the first amplification step - each photoactivated rhodopsin triggers activation of about 100 transducins. (The shape change in the opsin activates a G protein called transducin.)

— Each transducin then activates the enzyme cGMP-specific phosphodiesterase (PDE). (Transducin, in turn, activates the enzyme phosphodiesterase.)

— PDE then catalyzes the hydrolysis of cGMP. This is the second amplification step, where a single PDE hydrolyses about 1000 cGMP molecules. (The enzyme hydrolyzes the second messenger cGMP to GMP)

— With the intracellular concentration of cGMP reduced, the net result is closing of cyclic nucleotide-gated ion channels in the photoreceptor membrane because cGMP was keeping the channels open. (Because cGMP acts to keep Na+ ion channels open, the conversion of cGMP to GMP closes the channels.)

— As a result, sodium ions can no longer enter the cell, and the photoreceptor hyperpolarizes (its charge inside the membrane becomes more negative). (The closing of Na+ channels hyperpolarizes the cell.)

— This hyperpolarization means that less glutamate is released to the bipolar cell than before (see below). (The hyperpolarization of the cell slows the release of the neurotransmitter glutamate, which can either excite or inhibit the postsynaptic bipolar cells.)

— Reduction in the release of glutamate means one population of bipolar cells will be depolarized and a separate population of bipolar cells will be hyperpolarized, depending on the nature of receptors (ionotropic or metabotropic) in the postsynaptic terminal (see receptive field).

Thus, a rod or cone photoreceptor actually releases less neurotransmitter when stimulated by light.

ATP provided by the inner segment powers the sodium-potassium pump. This pump is necessary to reset the initial state of the outer segment by taking the sodium ions that are entering the cell and pumping them back out.

Although photoreceptors are neurons, they do not conduct action potentials with the exception of the ganglion cell photoreceptor.

Advantages

Phototransduction in rods and cones is unique in that the stimulus (in this case, light) actually reduces the cell's response or firing rate, which is unusual for a sensory system where the stimulus usually increases the cell's response or firing rate. However, this system offers several key advantages.

First, the classic (rod or cone) photoreceptor is depolarized in the dark, which means many sodium ions are flowing into the cell. Thus, the random opening or closing of sodium channels will not affect the membrane potential of the cell; only the closing of a large amount of channels, through absorption of a photon, will affect it and signal that light is in the visual field. Hence, the system is noiseless.

Second, there is a lot of amplification in two stages of classic phototransduction: one pigment will activate many molecules of transducin, and one PDE will cleave many cGMPs. This amplification means that even the absorption of one photon will affect membrane potential and signal to the brain that light is in the visual field. This is the main feature which differentiates rod photoreceptors from cone photoreceptors. Rods are extremely sensitive and have the capacity of registering a single photon of light unlike cones. On the other hand, cones are known to have very fast kinetics in terms of rate of amplification of phototransduction unlike rods.

Function

Photoreceptors do not signal color; they only signal the presence of light in the visual field.

A given photoreceptor responds to both the wavelength and intensity of a light source. For example, red light at a certain intensity can produce the same exact response in a photoreceptor as green light of a different intensity. Therefore, the response of a single photoreceptor is ambiguous when it comes to colour.

To determine color, the visual system compares responses across a population of photoreceptors (specifically, the three different cones with differing absorption spectra). To determine intensity, the visual system computes how many photoreceptors are responding. This is the mechanism that allows trichromatic color vision in humans and some other animals.

SIGNALLING

The rod and cone photoreceptors signal their absorption of photons through a release of the neurotransmitter glutamate to bipolar cells at its axon terminal. Since the photoreceptor is depolarized in the dark, a high amount of glutamate is being released to bipolar cells in the dark. Absorption of a photon will hyperpolarize the photoreceptor and therefore result in the release of *less* glutamate at the presynaptic terminal to the bipolar cell.

Every rod or cone photoreceptor releases the same neurotransmitter, glutamate. However, the effect of glutamate differs in the bipolar cells, depending upon the type of receptor imbedded in that cell's membrane. When glutamate binds to an ionotropic receptor, the bipolar cell will depolarize (and therefore will hyperpolarize with light as less glutamate is released). On the other hand, binding of glutamate to a metabotropic receptor results in a hyperpolarization, so this bipolar cell will depolarize to light as less glutamate is released.

In essence, this property allows for one population of bipolar cells that gets excited by light and another population that gets inhibited by it, even though all photoreceptors show the same response to light. This complexity becomes both important and necessary for detecting color, contrast, edges, etc.

Further complexity arises from the various interconnections among bipolar cells, horizontal cells, and amacrine cells in the retina. The final result is differing populations of ganglion cells in the retina, a sub-population of which is also intrinsically photosensitive, using the photopigment melanopsin.

GANGLION CELL (NON-ROD NON-CONE) PHOTORECEPTORS

In 1991 a non-rod non-cone photoreceptor in the eyes of mice was discovered where it was shown to mediate circadian rhythms i.e. the body's 24-hour biological clock. These novel cells express the photopigment melanopsin which was first identified by Ignacio Provencio and colleagues. Conclusively that cells containing the photopigment melanopsin absorb light maximally at different wavelength than those of rods and cones. In mice the non-rod non-cone photoreceptor had a role in initiating the pupil light reflex and not only circadian/behavioural functions as previously thought, though the latter were also demonstrated by them using genetically engineered rodless coneless mice. It is showed that in the rat intrinsically photosensitive retinal ganglion cells invariably expressed melanopsin and so melanopsin (and not rod or cone opsins) was most likely the visual pigment of phototransducing retinal ganglion cells that set the circadian clock and initiated other non-image-forming visual functions. This was highly significant anatomically - ganglion cells reside in the inner retina, while classic photoreceptors (rods and cones) inhabit the outer retina, suggesting two parallel and anatomically distinct photoreceptor pathways. It is showed (2005) that the melanopsin photopigment was the phototransduction pigment in ganglion cells. Dennis Dacey with colleagues showed in a species of Old World monkey that giant ganglion cells expressing melanopsin projected to the lateral geniculate nucleas. Previously only projections to the midbrain (pre-tectal nucleas) and hypothalamus (supra-chiasmatic nucleas) had been shown. However a visual role for the receptor was still unsuspected and unproven.

In 2007 researchers published the pioneering work using rodless coneless humans. Current Biology subsequently announced in their 2008 editorial, commentary and despatches to scientists and ophthalmologists, that the non-rod non-cone photoreceptor had been conclusively discovered in humans using landmark experiments on rodless coneless humans. The workers found the identity of the non-rod non-cone photoreceptor in humans to be a ganglion cell in the inner retina as had been shown previously in rodless coneless models in some other mammals. The workers had tracked down patients with rare diseases wiping out classic rod and cone photoreceptor function but preserving ganglion cell function. Despite having no rods or cones the patients continued to exhibit circadian photoentrainment, circadian behavioural patterns, melanopsin suppression, and pupil reactions, with peak spectral sensitivities to environmental and experimental light matching that for the melanopsin photopigment. Their brains could also associate vision with light of this frequency.

The use of rodless coneless humans allowed another possible role for the receptor to be studied. In 2007 a novel role was found for the photoreceptive ganglion cell. Researchers that the retinal ganglion cell was a photoreceptor (at least in humans) for conscious sight and not only for non-image-forming functions like circadian rhythms, behaviour and pupil reactions as previously thought. Humans were the perfect model in which to prove this function as they can describe sight readily to an observer which animals cannot do. Hence the receptor by its location anatomically in the inner retina as shown by these workers was the first cell to perceive light giving rise to vision. They also showed it responded mostly to blue light, suggesting it may have a role in mesopic vision. The rodless coneless human subjects hence also opened the door into image-forming (visual) roles for the ganglion cell photoreceptor. It was discovered that there are parallel pathways for vision - one classic rod and cone-based arising from the outer retina, the other a rudimentary visual brightness detector arising from the inner retina and which seems to be activated by light before the other. Classic photoreceptors also feed into the novel photoreceptor system, and colour constancy may be an important role as suggested by Foster. The receptor could be instrumental in understanding many diseases including major causes of blindness worldwide like glaucoma, a disease which affects ganglion cells, and the study of the receptor offered potential as a new avenue to explore in trying to find treatments for blindness. It is in these discoveries of the novel photoreceptor in humans and in the receptors role in vision, rather than its non-image-forming functions, where the receptor may have the greatest impact on society as a whole, though the impact of disturbed circadian rhythms is another area of relevance to clinical medicine.

Most work suggests that the peak spectral sensitivity of the receptor is between 460 and 481nm, though a minority of groups reported it being lower as far as 420nm. In 2003 researchers showed that 460 nm wavelengths of light suppress melatonin twice as much as longer 555 nm light. Using rodless coneless human, it was found that what consciously led to light perception was a very intense 481nm stimuli - this means that the receptor in visual terms enables some rudimentary vision maximally for blue light.

Photosensitive Ganglion Cell

Photosensitive ganglion cells, or melanopsin-containing ganglion cells, are a recently discovered type of nerve cell in the retina of the mammalian eye which, unlike other retinal ganglion cells, are intrinsically photosensitive. This means that they are a third class of retinal photoreceptors, excited by light even when all influences from classical photoreceptors (rods and cones) are blocked (either by applying pharmacological agents or by dissociating the ganglion cell from the retina). Intrinsically photosensitive retinal ganglion cells (ipRGC) contain the photopigment melanopsin. The giant retinal ganglion cells of the primate retina are examples of photosensitive ganglion cells.

Brief Overview

Compared to the rods and cones, the ipRGC respond more sluggishly and signal the presence of light over the long term. Their functional roles are non-image-forming and fundamentally different from those of pattern vision; they provide a stable representation of ambient light intensity. They have at least three primary functions.

- They play a major role in synchronizing circadian rhythms to the rising and setting of the sun. They send light information via the retinohypothalamic tract directly to the circadian pacemaker of the brain, the suprachiasmatic nucleus of the hypothalamus. The physiological properties of these ganglion cells match known properties of the light entrainment mechanism (synchronization) regulating circadian rhythms. This is the mechanism that allows us to overcome jet lag.
- Photosensitive ganglion cells also innervate other brain targets, such as the center of pupillary control, the olivary pretectal nucleus of the midbrain. They contribute to the regulation of pupil size and other behavioral responses to ambient lighting conditions.
- They contribute to photic regulation of, and acute photic suppression of, release of the hormone melatonin from the pineal gland.

Photosensitive ganglion cells are responsible for the persistence of circadian and pupillary light responses in mammals with degenerated rod and cone photoreceptors, such as humans suffering from retinitis pigmentosa.

Recently photoreceptive ganglion cells have been isolated in humans where, in addition to the above functions shown in other mammals, they have been shown to mediate a degree of rudimentary sight in rodless, coneless subjects suffering with disorders of rod and cone photoreceptors. The photoreceptive ganglion cells may have a visual function and can be isolated in humans.

The photopigment of these photoreceptive ganglion cells, melanopsin, absorbs light mainly in the blue portion of the visible spectrum (best wavelength = 480 nanometers). The phototransduction mechanism in these cells is not fully understood, but seems likely to resemble that in invertebrate rhabdomeric photoreceptors. Photosensitive ganglion cells respond to light by depolarizing and increasing the rate at which they fire nerve impulses. In addition to responding directly to light, these cells appear to receive excitatory and inhibitory influences from rods and cones by way of synaptic connections in the retina.

DISCOVERY OF PHOTORECEPTIVE GANGLION CELLS

In 1991 discovered a non-rod, non-cone photoreceptor in the eyes of mice where it was shown to mediate circadian rhythms i.e. the body's 24-hour biological clock The fact that such a landmark discovery was published in a relatively obscure science journal indicates the initial skepticism about the existence of non-rod, non-cone photoreceptors within the scientific community, which continued to widely believe that the only photoreceptors were rods and cones as if this were written in stone-and why not, after all, as Foster himself notes, the eye had been the subject of detailed study for a continuous period of over 200 years, so at the time it seemed unlikely that great minds since Newton, Maxwell, through to Einstein and beyond, could have missed this receptor's existence, its functions, and its ramifications. But miss it they did and it fell to contemporary researchers to make the landmark discoveries in the field, ground-breaking discoveries that still continue to be made. These novel cells express the photopigment melanopsin which was first identified by Ignacio Provencio and colleagues who published in the Journal of *Neuroscience* in 2000 Note how after almost one whole decade, major advances in the field would henceforth only be published in major biology and science journals, reflecting the gradual acceptance of the novel receptor by the scientific community.

Melanopsin Absorbs Different Maximal Wavelength

The cells containing the photopigment melanopsin absorb light maximally at different wavelength than those of rods and cones. Researchers also discovered that in mice the non-rod, non-cone photoreceptor had a role in initiating the pupil light reflex and not only circadian/behavioural functions as previously thought, though the latter were also demonstrated by them using genetically engineered rodless, coneless mice. Worker showed that in the rat, intrinsically photosensitive retinal ganglion cells invariably expressed melanopsin, and so melanopsin (and not rod or cone opsins) was most likely the visual pigment of phototransducing retinal ganglion cells that set the circadian clock and initiated other non-image-forming visual functions. This work is regarded by Current Biology, New Scientist and various other commentators as representing the discovery that the identity of the non-rod, non-cone photoreceptor in mice was a class of retinal ganglion cells (RGCs). This was highly significant anatomically - ganglion cells reside in the inner retina, while classic photoreceptors (rods and cones) inhabit the outer retina, suggesting two parallel and anatomically distinct photoreceptor pathways.

The melanopsin photopigment was the phototransduction pigment in ganglion cells (2005). A species of Old World monkey that giant ganglion cells expressing melanopsin projected to the lateral geniculate nucleas. Previously only projections to the midbrain (pre-tectal nucleas) and hypothalamus (supra-chiasmatic nucleas, SCN) had been shown. However a visual role for the receptor was still unsuspected and unproven.

Research in Humans

Attempts started to be made to hunt down the receptor in humans. But humans posed special challenges and demanded a new model - for unlike in animals, extensive ethical issues meant rod and cone loss could not be induced genetically or with chemicals so as to directly study the ganglion cells. For many years, only inferences could be drawn about the receptor in humans, though these were at times pertinent.

Current Biology subsequently announced in their 2008 editorial, commentary and despatches to scientists and ophthalmologists, that the non-rod, non-cone photoreceptor had been conclusively discovered in humans using landmark experiments on rodless, coneless humans. The 2007 discovery of the novel receptor in humans, as well as the spectacular discovery, made alongside, that it mediated conscious sight, was trumpeted by Cell Press, New Scientist, and other science commentators in 2007. The workers found the identity of the non-rod, non-cone photoreceptor in humans to be a ganglion cell in the inner retina as had been shown previously in rodless, coneless models in some other mammals. The workers had tracked down patients with rare diseases wiping out classic rod and cone photoreceptor function but preserving ganglion cell function. Despite having no rods or cones the patients continued to exhibit circadian photoentrainment, circadian behavioural patterns, melanopsin suppression, and pupil reactions, with peak spectral sensitivities to environmental and experimental light matching that for the melanopsin photopigment. Their brains could also associate vision with light of this frequency. Jacob Schor comments that in addition to being an outstanding example of collaboration between different countries, as well as between clinicians and scientists, interest thenceforth started to be shown by clinicians including ophthalmologists with a view to understanding the new receptor's role in human diseases.

New Role for Conscious Sight

The use of rodless, coneless humans allowed another possible role for the receptor to be studied. In 2007, it is showed that the retinal ganglion cell was a photoreceptor (at least in humans) for conscious sight and not only non-image-forming functions like circadian rhythms, behaviour and pupil reactions as previously thought. Humans were the perfect model in which to prove this function as they can describe sight readily to an observer, which animals cannot do. Hence the receptor by its location anatomically in the inner retina as shown by these researchers was the first cell to perceive light giving rise to vision. They also showed it responded most to blue light, suggesting it may have a role in mesopic vision and the old theory of a purely duplex retina with rod (dark) and cone (light) light vision was simplistic. Scientists work with rodless, coneless human subjects also opened the door into image-forming (visual) roles for the ganglion cell photoreceptor.

It also made the important discovery that there are parallel pathways for vision - one classic rod and cone-based arising from the outer retina, the other a rudimentary visual brightness detector arising from the inner retina and which seems to be activated by light before the other. lassic photoreceptors also feed into the novel photoreceptor system, and colour constancy may be an important role. Like many of the key discoveries about the new receptor, worker shatters hundreds of years of what science thought it knew about the most basic function of the eye and vision.

The authors on the rodless, coneless human model summarised their landmark paper noting for the first time that the receptor could be instrumental in understanding many diseases including major causes of blindness worldwide such as glaucoma, a disease which affects ganglion cells. Study of the receptor offered potential as a new avenue to explore in trying to find treatments for blindness. It is in these discoveries of the novel photoreceptor in humans and in the receptor's role in vision, rather than its non-image-forming functions, where the receptor may have the greatest impact on society as a whole, though the impact of disturbed circadian rhythms is another area of relevance to clinical medicine.

Violet-to-blue Light

Most work suggests that the peak spectral sensitivity of the receptor is between 460 and 484nm, though a minority of groups reported it being lower, as far as 420nm. Workers (2003) showed that 460nm (violet) wavelengths of light suppress melatonin twice as much as longer 555nm (green) light, the peak sensitivity of the photopic visual system. Workers using a rodless, coneless human, it was found that what consciously led to light perception was a very intense 481nm stimulus - this means that the receptor in visual terms enables some rudimentary vision maximally for blue light A potential criticism that the responses could have been due to heat would be missplaced, as heat is dissipated at higher wavelengths and would cause the sensation of greatest response with long wavelength (yellow and red) light, and not with short wavelength blue light as the researchers found.

G PROTEIN-COUPLED RECEPTOR

G protein-coupled receptors (GPCRs), also known as seven transmembrane domain receptors, 7TM receptors, heptahelical receptors, and G protein-linked receptors (GPLR), comprise a large protein family of transmembrane receptors that sense molecules outside the cell and activate inside signal transduction pathways and, ultimately, cellular responses. G protein-coupled receptors are found only in eukaryotes, including yeast, plants, choanoflagellates and animals. The ligands that bind and activate these receptors include light-sensitive compounds, odors, pheromones, hormones, and neurotransmitters, and vary in size from small molecules to peptides to large proteins. G protein-coupled receptors (GPCRs) are involved in many diseases, and are also the target of around half of all modern medicinal drugs.

Classification

GPCRs can be grouped into six classes based on sequence homology and functional similarity:

- Class A (or 1) (Rhodopsin-like)
- Class B (or 2) (Secretin receptor family)
- Class C (or 3) (Metabotropic glutamate/pheromone)
- Class D (or 4) (Fungal mating pheromone receptors)
- Class E (or 5) (Cyclic AMP receptors)
- Class F (or 6) (Frizzled/Smoothened)

The very large rhodopsin A group has been further subdivided into 19 subgroups (A1-A19). More recently, an alternative classification system called GRAFS (Glutamate, Rhodopsin, Adhesion, Frizzled/Taste2, Secretin) has been proposed.

The human genomes encodes roughly 350 G protein-coupled receptors, which detect hormones, growth factors and other endogenous ligands. Approximately 150 of the GPCRs found in the human genome have unknown functions.

Physiological Roles

The GPCRs are involved in a wide variety of physiological processes. Some examples of their physiological roles include:

— *The visual sense:* the opsins use a photoisomerization reaction to translate electromagnetic radiation into cellular signals. Rhodopsin, for example, uses the conversion of *11-cis*-retinal to *all-trans*-retinal for this purpose;

— *The sense of smell:* receptors of the olfactory epithelium bind odorants (olfactory receptors) and pheromones (vomeronasal receptors);

— *Behavioural and mood regulation:* receptors in the mammalian brain bind several different neurotransmitters, including serotonin, dopamine, GABA, and glutamate.

— *Regulation of immune system activity and inflammation:* chemokine receptors bind ligands that mediate intercellular communication between cells of the immune system; receptors such as histamine receptors bind inflammatory mediators and engage target cell types in the inflammatory response.

— *Autonomic nervous system transmission:* both the sympathetic and parasympathetic nervous systems are regulated by GPCR pathways, responsible for control of many automatic functions of the body such as blood pressure, heart rate, and digestive processes.

— *Cell density sensing:* A novel GPCR role in regulating cell density sensing.

Receptor Structure

The GPCRs are integral membrane proteins that possess seven membrane-spanning domains or transmembrane helices. The extracellular parts of the receptor can be glycosylated. These extracellular loops also contain two highly-conserved cysteine residues that form disulfide bonds to stabilize the receptor structure. Some seven transmembrane helix proteins (such as channelrhodopsin) that resemble GPCRs may contain different functional groups, such as entire ion channels, within their protein.

Early structural models for GPCRs were based on their weak analogy to bacteriorhodopsin for which a structure had been determined by both electron diffraction (PDB 2BRD, 1AT9) [and X ray-based crystallography (1AP9). In 2000, the first crystal structure of a mammalian GPCR, that of bovine rhodopsin (1F88), was solved While the main feature, the seven transmembrane helices, is conserved, the relative orientation of the helices differ significantly from that of bacteriorhodopsin. In 2007, the first structure of a human GPCR was solved (2R4R, 2R4S). This was followed immediately by a higher resolution structure of the same receptor (2RH1) This human $ß_2$-adrenergic receptor GPCR structure, proved to be highly similar to the bovine rhodopsin in terms of the relative orientation of the seven transmembrane helices. However the conformation of the second extracellular loop is entirely different between the two structures. Since this loop constitutes the "lid" that covers the top of the ligand binding site, this conformational difference highlights the difficulties in constructing homology models of other GPCRs based only on the rhodopsin structure.

Mechanism

G protein-coupled receptor are activated by an external signal in the form of a ligand or other signal mediator. This creates a conformational change in the receptor, causing activation of a G protein. Further effect depends on the type of G protein.

Ligand Binding

The GPCRs include receptors for sensory signal mediators (e.g., light and olfactory stimulatory molecules); adenosine, bombesin, bradykinin, endothelin, ?-aminobutyric acid (GABA), hepatocyte growth factor, melanocortins, neuropeptide Y, opioid peptides, opsins, somatostatin, tachykinins, vasoactive intestinal polypeptide family, and vasopressin; biogenic amines (e.g., dopamine, epinephrine, norepinephrine, histamine, glutamate (metabotropic effect), glucagon, acetylcholine (muscarinic effect), and serotonin); chemokines; lipid mediators of inflammation (e.g., prostaglandins, prostanoids, platelet-activating factor, and leukotrienes); and peptide hormones (e.g., calcitonin, C5a anaphylatoxin, follicle-stimulating hormone (FSH), gonadotropic-releasing hormone (GnRH), neurokinin, thyrotropin-releasing hormone (TRH), and oxytocin). GPCRs that act as receptors for stimuli that have not yet been identified are known as orphan receptors.

Whereas, in other types of receptors that have been studied, ligands bind externally to the membrane, the ligands of GPCRs typically bind within the transmembrane domain.

Conformational Change

The transduction of the signal through the membrane by the receptor is not completely understood. It is known that the inactive G protein is bound to the receptor in its inactive state. Once the ligand is recognized, the receptor shifts conformation and thus mechanically activates the G protein, which detaches from the receptor. The receptor can now either activate another G protein or switch back to its inactive state. This is an overly simplistic explanation, but suffices to convey the overall set of events.

It is believed that a receptor molecule exists in a conformational equilibrium between active and inactive biophysical states. The binding of ligands to the receptor may shift the equilibrium toward the active receptor states. Three types of ligands exist: agonists are ligands that shift the equilibrium in favour of active states; inverse agonists are ligands that shift the equilibrium in favour of inactive states; and neutral antagonists are ligands that do not affect the equilibrium. It is not yet known how exactly the active and inactive states differ from each other.

Activation of G Protein

If a receptor in an active state encounters a G protein, it may activate it. Some evidence suggests that receptors and G proteins are actually pre-coupled. For example, binding of G proteins to receptors affects the receptor's affinity for ligands. Activated G proteins are bound to GTP.

Further signal transduction depends on the type of G protein. The enzyme adenylate cyclase is an example of a cellular protein that can be regulated by a G protein, in this case the G protein Gs. Adenylate cyclase activity is activated when it binds to a subunit of the activated G protein. Activation of adenylate cyclase ends when the G protein returns to the GDP-bound state.

GPCR Signaling without G Proteins

In the late 1990s, evidence began accumulating to suggest that some GPCRs are able to signal without G proteins. The ERK2 mitogen-activated protein kinase, a key signal transduction mediator downstream of receptor activation in many pathways, has been shown to be activated in response to cAMP-mediated receptor activation in the slime mold *D. discoideum* despite the absence of the associated G protein a- and ß-subunits.

In mammalian cells, the much-studied ß2-adrenoceptor has been demonstrated to activate the ERK2 pathway after arrestin-mediated uncoupling of G-protein-mediated signaling. It therefore seems likely that some mechanisms previously believed to be purely related to receptor desensitisation are actually examples of receptors switching their signaling pathway rather than simply being switched off.

In kidney cells, the bradykinin receptor B2 has been shown to interact directly with a protein tyrosine phosphatase. The presence of a tyrosine-phosphorylated ITIM (immunoreceptor tyrosine-based inhibitory motif) sequence in the B2 receptor is necessary to mediate this interaction and subsequently the antiproliferative effect of bradykinin

Receptor Regulation

The GPCRs become desensitized when exposed to their ligand for a prolonged period of time. There are two recognized forms of desensitization: 1) homologous desensitization, in which the activated GPCR is downregulated; and 2) heterologous desensitization, wherein the activated GPCR causes downregulation of a different GPCR. The key reaction of this downregulation is the phosphorylation of the intracellular (or cytoplasmic) receptor domain by protein kinases.

Phosphorylation by cAMP-dependent Protein Kinases

Cyclic AMP-dependent protein kinases (protein kinase A) are activated by the signal chain coming from the G protein (that was activated by the receptor) via adenylate cyclase and cyclic AMP (cAMP). In a *feedback mechanism*, these activated kinases phosphorylate the receptor. The longer the receptor remains active, the more kinases are activated, the more receptors are phosphorylated. In $ß_2$-adrenoceptors, this phosphorylation results in the switching of the coupling from the G_0 class of G-protein to the G_i class. cAMP-dependent PKA mediated phosphorylation is also known as heterologous desensitisation, because it is not specific to ligand bound receptor. In fact any receptor causing an increase in PKA activity will cause increased amounts of this type of desensitisation of other receptors coupled to G_0 (*e.g.*, dopamine receptor D_2 activation may lead to $ß_2$-adrenoceptor desensitisation of this type).

Phosphorylation by GRKs

The G protein-coupled receptor kinases (GRKs) are protein kinases that phosphorylate only active GPCRs.

Phosphorylation of the receptor can have two consequences:

Translocation: The receptor is, along with the part of the membrane it is embedded in, brought to the inside of the cell, where it is dephosphorylated within the acidic vesicular environment and then brought back. This mechanism is used to regulate long-term exposure, for example, to a hormone, by allowing resensitisation to follow desensitisation. Alternatively, the receptor may undergo lysozomal degradation, or remain internalised, where it is thought to participate in the initiation of signalling events, the nature of which depend on the internalised vesicle's subcellular localisation.

Arrestin linking: The phosphorylated receptor can be linked to *arrestin* molecules that prevent it from binding (and activating) G proteins, effectively switching it off for a short period of time. This mechanism is used, for example, with rhodopsin in retina cells to compensate for exposure to bright light. In many cases, arrestin binding to the receptor is a prerequisite for translocation. For example, beta-arrestin bound to $ß_2$-adrenoreceptors acts as an adaptor for binding with clathrin, and with the beta-subunit of AP_2 (clathrin adaptor molecules); thus the arrestin here acts as a scaffold assembling the componenets needed for clathrin-mediated endocytosis of $ß_2$-adrenoreceptors.

Receptor Oligomerization

It is generally accepted that G-protein-coupled receptors can form homo- and/or heterodimers and possibly more complex oligomeric structures, and indeed heterodimerization has been shown to be essential for the function of receptors such as the metabotropic GABA(B) receptors. However, it is presently unproven that true heterodimers exist. Present biochemical and physical techniques lack the resolution to differentiate between distinct homodimers assembled into an oligomer or true 1:1 heterodimers. It is also unclear what the functional significance of oligomerization might be, although it is thought that the phenomenon may contribute to the pharmacological heterogeneity of GPCRs in a manner not previously anticipated. This is an actively-studied area in GPCR research.

The best studied example of receptor oligomerisation are the metabotropic $GABA_B$ receptors. These receptors are formed by heterodimerization of $GABA_BR1$ and $GABA_BR2$ subunits. Expression of the $GABA_BR1$ without the $GABA_BR2$ in heterologous systems leads to retention of the subunit in the endoplasmic reticulum. Expression of the $GABA_BR2$ subunit alone, meanwhile, leads to surface expression of the subunit, although with no functional activity (*i.e.*, the receptor does not bind agonist and cannot initiate a response following exposure to agonist). Expression of the two subunits together leads to plasma membrane expression of functional receptor. It has been shown that $GABA_BR2$ binding to $GABA_BR1$ causes masking of a retention signal of functional receptors.

Plants

The GCR2 is a G-protein-coupled receptor for the plant hormone abscisic acid that has been identified in *Arabidopsis thaliana*. Another putative receptor is GCR1 for which no ligand has been identified yet.

Dictyostelium

A novel GPCR containing a lipid kinase domain has recently been identified in *Dictyostelium* that regulates cell density sensing.

13

Plant Pigments for Colour and Nutrition

INTRODUCTION

The pigments found in plants play important roles in plant metabolism and visual attraction in nature. They are also important for humans, attracting our attention and providing us with nutrients. Major plant pigments include carotenoids, anthocyanins and other flavonoids, betalains, and chlorophylls.

Chlorophylls, which are green, and carotenoids, which are yellow, orange or red, play pivotal roles in photosynthesis. They occur in all green plants and are localized in plastids. Chlorophylls capture light energy and convert it into chemical energy. They rarely occur outside of photosynthetic tissue. Carotenoids protect the chlorophylls from photo-oxidation and are accessory, light-harvesting pigments and photoreceptors. They occur in all green tissues as well as independently of chlorophyll in flowers (where they serve to attract animals), storage organs, and other plant parts.

Flavonoids include red or blue anthocyanins and white or pale yellow compounds such as rutin, quercitin, and kaempferol. Flavonoids in flowers and fruit provide visual cues for animal pollinators and seed dispersers to locate their target. They also occur in most other plant parts and in most genera. Flavonoids are located in the cytoplasm and plastids. Like carotenoids and flavonoids in flowers and fruit, betalains also are likely to play an important role in attracting animals. These red-violet (betacyanin) and yellow (betaxanthin) pigments, which are located in the cytoplasm of plant tissue, only occur in about 10 plant families (and always independent of anthocyanins).

Plant pigments are important cues to humans and other herbivorous animals in helping identify plants, find plant parts such as fruit, leaves, stems, roots, or tubers, and determine stages of plant development such as fruit ripeness or overall senescence. It was realized early in this century that many of these pigments play a positive role in human health. In 1919, Steenbock noted that yellow corn (*Zea mays* L.) and "yellow" vegetables (carrots (*Daucus carota* L.) and sweetpotato (*Ipomoea batatas* L.)) eliminated the symptoms of vitamin A deficiency in rats while white corn and "white" vegetables (parsnip (*Pastinoca sativa* L.), potato (*Solanum tuberosum* L.), and beets (*Beta vulgaris* L.) did not. Since then, approximately 40 carotenoids have been found to be vitamin A precursors (Simpson, 1983). When provitamin A carotenoids are consumed, they are enzymatically broken down to retinol (vitamin A). In this way, consumption of horticultural crops provides over 80% of the vitamin A for the world. Vitamin A deficiency worldwide is the most common specific dietary deficiency as it afflicts millions of children each year with xerophthalmia, blindness, or death. Subclinical deficiency also reduces immune function to increase the risk of severe and fatal infections. Therefore, the development of new and more potent sources of provitamin A carotenoids in horticultural crops, and improvement of production, shelf life and consumer acceptance of these crops can make an important contribution to improved human health.

Research findings since the 1970s have indicated that food crops containing carotenoids, anthocyanins, and other flavonoids are believed to function as "chemopreventers", by providing protection against certain forms of cancer and a reduction of cardiovascular disease. Interestingly, non-provitamin A carotenoids such as lycopene, the major pigment of tomato (*Lycopersicon esculentum* L. and watermelon (*Citrullus lanatus*

Thunb.), also confer chemoprevention. The important characteristic common to all chemopreventers is their antioxidant capacity. The antioxidant properties of carotenoids which protect plants in photosynthesis apparently may also protect humans from carcinogens and heart disease. Plant pigments are only a subset of the naturally occurring chemopreventers in vegetables and fruits. Other chemopreventive antioxidants in plants include vitamin C, vitamin E, phenolic acids, and organosulfur compounds.

While many clinical trials supplementing the diet with individual antioxidants have been successful in reducing health risks, some trials have not, for reasons yet unclear. Because of the complexity of large scale efficacy testing of natural compounds in foods, dietary supplementation with plant pigments and other chemopreventers has not yet been undertaken. However increased intake of the best sources of these compounds - fruits and vegetables - is encouraged.

Horticulturists have long studied plant pigments in vegetables, fruits and ornamental crops because of their vital role in visual appeal. More recently efforts have included nutritional evaluation and improvement of food crops as a source of vitamins. The orange and red vegetables (carrot, sweetpotato, and tomato) as major sources of provitamin A carotenoids. Compared to typical carrots, sweetpotatoes, and tomatoes which contain 160, 120, and 7 mg.kg-1 (fresh weight) of provitamin A carotenoids, respectively, high-carotene genetic sources containing 600, 170, and 95 mg.kg-1, respectively, have been bred (Simon, 1990, 1992; Simon et al., 1989; also see cover). Even typically low-carotene vegetables, such as cauliflower (*Brassica oleracea*) (Dickson et al., 1988) and cucumber (*Cucumis sativus* L.) can be genetically improved to serve as a source of carotenoids.

With the increasing evidence of the positive influence of plant pigments on human health, a heightened interest in a broad range of vegetable pigments and chemopreventers is developing. Genetic improvement of anthocyanin or betalain content has been successful for a variety of vegetables. Chemical and biotechnological approaches to improvement of plant pigments are also encouraging.

Horticultural approaches for improving the nutritional quality and visual appeal of our food supply provide a sustainable, inexpensive complement to medical and social programs for preventing human disease. Plant pigments provide a common interface familiar to horticultural researchers, medical researchers, students, producers, marketers, and consumers. Discussions at this interface will relate the relevance of science, in general, and horticulture, in particular, to the general public and can stimulate student interest.

PIGMENTS ARE COLOURFUL COMPOUNDS

Pigments are chemical compounds which reflect only certain wavelengths of visible light. This makes them appear "colourful". Flowers, corals, and even animal skin contain pigments which give them their colors. More important than their reflection of light is the ability of pigments to absorb certain wavelengths.

Because they interact with light to absorb only certain wavelengths, pigments are useful to plants and other autotrophs —organisms which make their own food using photosynthesis. In plants, algae, and cyanobacteria, pigments are the means by which the energy of sunlight is captured for photosynthesis. However, since each pigment reacts with only a narrow range of the spectrum, there is usually a need to produce several kinds of pigments, each of a different color, to capture more of the sun's energy.

There are three basic classes of pigments:

Chlorophylls are greenish pigments which contain a porphyrin ring. This is a stable ring-shaped molecule around which electrons are free to migrate. Because the electrons move freely, the ring has the potential to gain or lose electrons easily, and thus the potential to provide energized electrons to other molecules. This is the fundamental process by which chlorophyll "captures" the energy of sunlight.

There are several kinds of chlorophyll, the most important being chlorophyll "a". This is the molecule which makes photosynthesis possible, by passing its energized electrons on to molecules which will manufacture sugars. All plants, algae, and cyanobacteria which photosynthesize contain chlorophyll "a". A second kind of chlorophyll is chlorophyll "b", which occurs only in "green algae" and in the plants. A third form of chlorophyll which is common is (not surprisingly) called chlorophyll "c", and is found only in the photosynthetic members of the Chromista as well as the dinoflagellates. The differences between the chlorophylls of these major groups was one of the first clues that they were not as closely related as previously thought.

Carotenoids are usually red, orange, or yellow pigments, and include the familiar compound carotene, which gives carrots their colour. These compounds are composed of two small six-carbon rings connected by a "chain" of carbon atoms. As a result, they do not dissolve in water, and must be attached to membranes within the cell. Carotenoids cannot transfer sunlight energy directly to the photosynthetic pathway, but must pass their absorbed energy to chlorophyll. For this reason, they are called accessory pigments. One very visible accessory pigment is fucoxanthin the brown pigment which colors kelps and other brown algae as well as the diatoms.

Phycobilins are water-soluble pigments, and are therefore found in the cytoplasm, or in the stroma of the chloroplast. They occur only in Cyanobacteria and Rhodophyta.

The picture at the right shows the two classes of phycobilins which may be extracted from these "algae". The vial on the left contains the bluish pigment phycocyanin, which gives the Cyanobacteria their name. The vial on the right contains the reddish pigment phycoerythrin, which gives the red algae their common name.

Phycobilins are not only useful to the organisms which use them for soaking up light energy; they have also found use as research tools. Both pycocyanin and phycoerythrin fluoresce at a particular wavelength. That is, when they are exposed to strong light, they absorb the light energy, and release it by emitting light of a very narrow range of wavelengths. The light produced by this fluorescence is so distinctive and reliable, that phycobilins may be used as chemical "tags". The pigments are chemically bonded to antibodies, which are then put into a solution of cells. When the solution is sprayed as a stream of fine droplets past a laser and computer sensor, a machine can identify whether the cells in the droplets have been "tagged" by the antibodies. This has found extensive use in cancer research, for "tagging" tumor cells.

PIGMENTS

Pigments are "molecules that absorb specific wavelengths (energies) of light and reflect all others."

Pigments are coloured: the colour we see is the net effect of all the light reflecting back at us.

Electrons in molecules can exist at specific energy levels. Normally they exist at the lowest possible energy level they can. However, if enough energy comes along to boost them into the next level, they can "absorb" that energy and occupy that higher level. This is what pigments do. The light they absorb contains' just the right amount' of energy necessary to push them into the next level. Any light that does not have enough or has too much energy can not be absorbed and is reflected.

The electron in the higher energy level, however, does not 'want' to stay there (i.e. it is unstable). It 'wants' to return to its normal lower energy level. In order to do this it must get rid or release the energy that has put it into the higher energy state to begin with. This can happen several different ways:

1. The extra energy can be converted into molecular motion and lost as heat.
2. Some of the extra energy can be lost as heat energy, while the rest is lost as light. This re-emission of light energy is called florescence.
3. The energy, but not the e-itself, can be passed onto another molecule. This is called resonance.
4. The energy and the e-can be transferred to another molecule.

Plant pigments usually utilize the last two of these reactions to convert the sun's energy into their own. When chlorophyll is isolated from the enzymes it is associated with, the second scenario can be seen to happen.

What should be the ideal pigment for chloroplasts?

A collection of pigments that would absorb all light and thus appear Black seems a logical choice . . . but in fact we know this is not true plants except for some of the red algae appear green or brown, not black. Why? a number of possible explanation occur . . .

IF plants had pigments that absorbed UV and *x*-rays this would mean that so much energy could be absorbed in light areas that electrons could be knocked off their orbitals and the molecule destroyed.

IF plants absorbed IR and radio waves, there would not be enough energy for electron transfer, just enough to warm up the molecule

Pigments that absorb in the visible region gain just enough energy to boost an electron to the next level.

However even in this region, not all visible wavelengths are abosrbed . . . there is some speculation that in the early competitive wars between photosynthetic bacteria [and we mean early 3 BYA] , plants specialized so that some would absorb in the red, others green or blue. The survivors in the long run are the one with pigments that absorbed in the red/blue reflecting green . . . at least on land.

What are these Pigments involved in Photosythesis?

- *Chlorophyll a:* This is the most abundant pigment in plants. Chlorophyll a absorbs light with wavelengths of 430nm(blue) and 662nm(red). It reflects green light strongly so it appears green to us. It contains a hydrophobic (fat soluble) phytol chain that allow it to be embedded in a lipid membrane. The rest of the structure called a tetrapyrrolic ring rests outside of the membrane . It is this part of the pigment that absorbs the energy from light. The metal at the center of the structure, Mg, can have variable oxidation states . This means that it can accept and donate e- readily depending of the situation. Its flexible, which is very important to the function of the molecule.

- *Chlorophyll b:* This molecule has a structure similar to that of chlorophyll a. It absorbs light of 453nm and 642 nm maximally. It is not as abundant as chlorophyll a, and probably evolved later. It helps increase the range of light a plant can use for energy.
- *Carotenoids:* This is a class of **accessory** pigments that occur in all photosynthetic organisms. They are completely hydrophobic (fat soluble) and exist in lipid membranes. Carotenoids absorb light maximally between 460 nm and 550 nm and appear red, orange, or yellow to us.

The most important function of carotenoids seems to be protecting the plant from free radicals formed from ultra violet or other radiation. Free radicals are dangerous because they contain an extra odd e-they don't really want to have. This means that they are constantly trying to get rid of this extra electron. They do this by attacking whatever bonds they can.

In animal systems it is speculated that:

- Vitamin E + radical oxidized $\rightarrow$ radical + Vitamin E oxidized
- Vitamin E oxidized + carotenoids $\rightarrow$ Vit E + carotenoids oxidized
- Carotenoids oxidized + Vit C $\rightarrow$ carotenoids + Vit C oxidized

In animals, the Vit C can be flushed out of the system since it is soluble in water which none of the other molecules are. Whether this same series of transfers occurs in plants is unknown. In humans, taking Vit E or betacarotenes in themselves is not protection against oxidants unless there is a supply of Vit C to flush it out. Smokes who maintain low levels of C are not helped with increases of the other 2.

In plants, as the vacuole is a generally safe aqueous repository, the oxidized C may end up here, whereas carotenoids could not.

- Xanthophylls are a fourth common class of pigments. They are essentially oxidized Carotenoids and contain oxygen. They are usually red and yellow and do not absorb energy as well as cartenoids. They are also fat soluble.
- Anthocyanins are a fifth class of pigments. These pigments contain Cu and are stored in the vacuole of a plant because they unlike the other fat soluble pigments are water soluble.
- Note the structure of the phycocyanins we in the red algae and BG's. How does it differ substantially from the above pigments?

There are many good reasons for knowing about plant pigments (and for producing a book about them). Plant pigments are important in signalling, as in attracting pollinating and dispersal agents, and repelling herbivores. Plant pigments are obviously physiologically important, and recent research suggests novel protective mechanisms, both photo-protective and anti-oxidative. Plant pigments are economically important, in determining the colours and patterns of attractive flowers and valuable fruits. They are also important nutritionally, with an understanding of emerging new roles in nutrition and in aiding health. With all of this interest, there is a surprising lack of scholarly books on plant pigments.

14

Chloroplast

INTRODUCTION

Chloroplasts are organelles found in plant cells and eukaryotic algae that conduct photosynthesis. Chloroplasts absorb light and use it in conjunction with water and carbon dioxide to produce sugars, the raw material for energy and biomass production in all green plants and the animals that depend on them, directly or indirectly, for food. Chloroplasts capture light energy to conserve free energy in the form of ATP and reduce NADP to NADPH through a complex set of processes called photosynthesis. The word chloroplast is derived from the Greek words *chloros* which means green and *plast* which means form or entity. Chloroplasts are members of a class of organelles known as plastids.

Chloroplasts are one of the many different types of organelles in the cell. They are generally considered to have originated as endosymbiotic cyanobacteria (i.e. blue-green algae). This was first suggested by Mereschkowsky in 1905 after an observation by Schimper in 1883 that chloroplasts closely resemble cyanobacteria. All chloroplasts are thought to derive directly or indirectly from a single endosymbiotic event (in the Archaeplastida), except for *Paulinella chromatophora*, which has recently acquired a photosynthetic cyanobacterial endosymbiont which is not closely related to chloroplasts of other eukaryotes. that they derive from an endosymbiotic event, chloroplasts are similar to mitochondria but chloroplasts are found only in plants and protista. The chloroplast is surrounded by a double-layered composite membrane with an intermembrane space; it has its own DNA and is involved in energy metabolism. Further, it has reticulations, or many infoldings, filling the inner spaces.

In green plants, chloroplasts are surrounded by two lipid-bilayer membranes. The inner membrane is now believed to correspond to the outer membrane of the ancestral cyano-bacterium. Chloroplasts have their own genome, which is considerably reduced compared to that of free-living cyanobacteria, but the parts that are still present show clear similarities with the cyanobacterial genome. Plastids may contain 60-100 genes whereas cyanobacteria often contain more than 1500 genes Many of the missing genes are encoded in the nuclear genome of the host. The transfer of nuclear information has been estimated in tobacco plants at one gene for every 16000 pollen grains.

In some algae (such as the heterokonts and other protists such as Euglenozoa and Cercozoa), chloroplasts seem to have evolved through a secondary event of endosymbiosis, in which a eukaryotic cell engulfed a second eukaryotic cell containing chloroplasts, forming chloroplasts with three or four membrane layers. In some cases, such secondary endosymbionts may have themselves been engulfed by still other eukaryotes, thus forming tertiary endosymbionts. In the alga *Chlorella*, there is only one chloroplast, which is bell shaped.

In some groups of mixotrophic protists such as the dinoflagellates, chloroplasts are separated from a captured alga or diatom and used temporarily. These klepto chloroplasts may only have a lifetime of a few days and are then replaced.

STRUCTURE

Chloroplasts are observable morphologically as flat discs usually 2 to 10 micrometer in diameter and 1 micrometer thick. In land plants they are generally 5 μm in diameter and 2.3 μm thick. The chloroplast is contained by an envelope that consists of an inner and an outer phospholipid membrane. Between these two layers is the intermembrane space. A typical [parenchyma] cell contains about 10 to 100 chloroplasts.

The material within the chloroplast is called the stroma, corresponding to the cytosol of the original bacterium, and contains one or more molecules of small circular DNA. It also contains ribosomes, although most of its proteins are encoded by genes contained in the host cell nucleus, with the protein products transported to the chloroplast.

Within the stroma are stacks of thylakoids, the sub-organelles which are the site of photosynthesis. The thylakoids are arranged in stacks called grana (singular: granum). A thylakoid has a flattened disk shape. Inside it is an empty area called the thylakoid space or lumen. Photosynthesis takes place on the thylakoid membrane; as in mitochondrial oxidative phosphorylation, it involves the coupling of cross-membrane fluxes with biosynthesis via the dissipation of a proton electrochemical gradient.

In the electron microscope, thylakoid membranes appear as alternating light-and-dark bands, each 0.01 μm thick. Embedded in the thylakoid membrane is the antenna complex, which consists of the light-absorbing pigments, including chlorophyll and carotenoids, and proteins (which bind the chlorophyll). This complex both increases the surface area for light capture, and allows capture of photons with a wider range of wavelengths. The energy of the incident photons is absorbed by the pigments and funneled to the reaction centre of this complex through resonance energy transfer. Two chlorophyll molecules are then ionised, producing an excited electron which then passes onto the photochemical reaction centre.

Recent studies have shown that chloroplasts can be interconnected by tubular bridges called stromules, formed as extensions of their outer membranes Chloroplasts appear to be able to exchange proteins via stromules and thus function as a network.

TRANSPLASTOMIC PLANTS

Recently, chloroplasts have caught attention by developers of genetically modified plants. In most flowering plants, chloroplasts are not inherited from the male parent although in plants such as pines, chloroplasts are inherited from males. Where chloroplasts are inherited only from the female, transgenes in these plastids cannot be disseminated by pollen. This makes plastid transformation a valuable tool for the creation and cultivation of genetically modified plants that are biologically contained, thus posing significantly lower environmental risks. This biological containment strategy is therefore suitable for establishing the coexistence of conventional and organic agriculture. The reliability of this mechanism has not yet been studied for all relevant crop species. However, the research programme published results for tobacco plants, demonstrating that the containment of transplastomic plants is highly reliable with a tiny failure rate of 3 in 1,000,000

Calvin Cycle

The Calvin cycle (or Calvin-Benson-Bassham cycle or carbon fixation) is a series of biochemical reactions that takes place in the stroma of chloroplasts in photosynthetic organisms. It was discovered by Melvin Calvin, James Bassham and Andrew Benson at the University of California, Berkeley It is one of the light-independent reactions or dark reactions.

Overview

During photosynthesis, light energy is used to generate chemical free energy, stored in glucose. The light-independent Calvin cycle, also (misleadingly) known as the "dark reaction" or "dark stage", uses the energy from short-lived electronically-excited carriers to convert carbon dioxide and water into organic compounds that can be used by the organism (and by animals which feed on it). This set of reactions is also called *carbon fixation*. The key enzyme of the cycle is called RuBisCO. In the following equations, the chemical species (phosphates and carboxylic acids) exist in equilibria among their various ionized states as governed by the pH.

The enzymes in the Calvin cycle are functionally equivalent to many enzymes used in other metabolic pathways such as gluconeogenesis and the pentose phosphate pathway, but they are to be found in the chloroplast stroma instead of the cell cytoplasm, separating the reactions. They are activated in the light (which is why the name "dark reaction" is misleading), and also by products of the light-dependent reaction. These regulatory functions prevent the Calvin cycle from operating in reverse to respiration, which would create a continuous cycle of carbon dioxide being reduced to carbohydrates, and carbohydrates being respired to carbon dioxide. Energy (in the form of ATP) would be wasted in carrying out these reactions that have no net productivity.

The sum of reactions in the Calvin cycle is the following:

$$3\ CO_2 + 6\ NADPH + 5\ H_2O + 9\ ATP\ ?\ C_3H_5O_3\text{-}PO_3^{2-} + 2\ H^+ + 6\ NADP^+ + 9\ ADP + 8\ P_i$$

OR

$$3\ CO_2 + 6\ C_{21}H_{29}N_7O_{17}P_3 + 5\ H_2O + 9\ C_{10}H_{16}N_5O_{13}P_3\ ?\ C_3H_5O_3 - PO_3^{2-} + 2\ H^+ + 6\ NADP^+ + 9\ C_{10}H_{15}N_5O_{10}P_2 + 8\ P_i$$

It should be noted that hexose (six carbon) sugars are not a product of the Calvin cycle. Although many texts list a product of photosynthesis as $C_6H_{12}O_6$, this is mainly a convenience to counter the equation of respiration, where six-carbon sugars are oxidized in mitochondria. The carbohydrate products of the Calvin Cycle are three-carbon sugar phosphate molecules, or "triose phosphates," specifically, glyceraldehyde-3-phosphate (G3P).

Steps of the Calvin Cycle

- The enzyme RuBisCO catalyses the carboxylation of Ribulose-1,5-bisphosphate, a 5 carbon compound, by carbon dioxide (a total of 6 carbons) in a two-step reaction. The initial product of the reaction is a six-carbon intermediate so unstable that it immediately splits in half, forming two molecules of glycerate 3-phosphate, a 3-carbon compound. (also: 3-phosphoglycerate, 3-phosphoglyceric acid, 3PGA)
- The enzyme phosphoglycerate kinase catalyses the phosphorylation of 3PGA by ATP (which was produced in the light-dependent stage). 1,3-bisphosphoglycerate (glycerate-1,3-bisphosphate) and ADP are the products.

(However, note that two PGAs are produced for every CO_2 that enters the cycle, so this step utilizes 2ATP per CO_2 fixed.

- The enzyme G3P dehydrogenase catalyses the reduction of 1,3BPGA by NADPH (which is another product of the light-dependent stage). Glyceraldehyde 3-phosphate (also G3P, GP) is produced, and the NADPH itself was oxidized and becomes NADP+. Again, two NADPH are utilized per CO_2 fixed.

(Simplified versions of the Calvin cycle integrate the remaining steps, except for the last one, into one general step - the regeneration of RuBP - also, one G3P would exit here.)

- Triose phosphate isomerase converts some G3P reversibly into dihydroxyacetone phosphate (DHAP), also a 3-carbon molecule.
- Aldolase and fructose-1,6-bisphosphatase convert a G3P and a DHAP into fructose-6-phosphate (6C). A phosphate ion is lost into solution.

  Then fixation of another CO_2 generates two more G3P.
- F6P has two carbons removed by transketolase, giving erythrose-4-phosphate. The two carbons on transketolase are added to a G3P, giving the ketose xylulose-5-phosphate (Xu5P).
- E4P and a DHAP (formed from one of the G3P from the second CO2 fixation) are converted into sedoheptulose-1,7-bisphosphate (7C) by aldolase enzyme.
- Sedoheptulose-1,7-bisphosphatase (one of only three enzymes of the Calvin cycle which are unique to plants) cleaves sedoheptulose-1,7-bisphosphate into sedoheptulose-7-phosphate, releasing an inorganic phosphate ion into solution.
- Fixation of a third CO_2 generates two more G3P. The ketose S7P has two carbons removed by transketolase, giving ribose-5-phosphate (R5P), and the two carbons remaining on transketolase are transferred to one of the G3P, giving another Xu5P. This leaves one G3P as the product of fixation

of 3 CO_2, with generation of three pentoses which can be converted to Ru5P.

R5P is converted into ribulose-5-phosphate (Ru5P, RuP) by phosphopentose isomerase. Xu5P is converted into RuP by phosphopentose epimerase.

Finally, phosphoribulokinase (another plant unique enzyme of the pathway) phosphorylates RuP into RuBP, ribulose-1,5-bisphosphate, completing the Calvin *cycle*. This requires the input of one ATP.

Thus, of 6 G3P produced, three RuBP (5C) are made totalling 15 carbons, with only one available for subsequent conversion to hexose. This required 9 ATPs and 6 NADPH per 3 CO_2.

RuBisCO also reacts competitively with O_2 instead of CO_2 in *photorespiration*. The rate of photorespiration is higher at high temperatures. "photorespiration" turns RuBP into 3PGA and 2-phosphoglycolate, a 2-carbon molecule which can be converted via glycolate and glyoxalate to glycine. Via the glycine cleavage system and tetrahydrofolate, two glycines are converted into serine +CO_2. Serine can be converted back to 3-phosphoglycerate. Thus, only 3 of 4 carbons from two phosphoglycolates can be converted back to 3PGA. Obviously photorespiration has very negative consequences for the plant, because rather than fixing CO_2, this process leads to loss of CO_2. C4 carbon fixation evolved to circumvent photorespiration, but can only occur in certain plants living in very warm or tropical climates.

PRODUCTS OF THE CALVIN CYCLE

The immediate product of the Calvin cycle is glyceraldehyde-3-phosphate (G3P) and water. Two G3P molecules (or one F6P molecule) that have exited the cycle are used to make larger carbohydrates. In simplified versions of the Calvin cycle they may be converted to F6P or F5P after exit, but this conversion is also part of the cycle.

Hexose isomerase converts about half of the F6P molecules in to glucose-6-phosphate. These are dephosphorylated and the glucose can be used to form starch, which is stored in, for example, potatoes, or cellulose used to build up cell walls. Glucose, with fructose, forms sucrose, a non-reducing sugar which is a stable storage sugar, unlike glucose.

C_4 Carbon Fixation

C_4 carbon fixation is one of three biochemical mechanisms, along with C_3 and CAM photosynthesis, functioning in land plants to "fix" carbon dioxide (binding the gaseous molecules to dissolved compounds inside the plant) for sugar production through photosynthesis. Along with CAM photosynthesis, C_4 fixation is considered an advancement over the simpler and more ancient C_3 carbon fixation mechanism operating in most plants. Both mechanisms overcome the tendency of RuBisCO (the first enzyme in the Calvin cycle) to photorespire, or waste energy by using oxygen to break down carbon compounds to CO_2. However C_4 fixation requires more energy input than C_3 in the form of ATP. C_4 plants separate rubisco from atmospheric oxygen, fixing carbon in the mesophyll cells and using oxaloacetate and malate to ferry the fixed carbon to rubisco and the rest of the Calvin cycle enzymes isolated in the bundle-sheath cells. The intermediate compounds both contain four carbon atoms, hence the name C_4.

The Pathway

The C_4 pathway was discovered by M.D. Hatch and C.R. Slack, in Australia, in 1966, so it is sometimes called the Hatch-Slack pathway.

In C_3 plants, the first step in the light-independent reactions of photosynthesis involves the fixation of CO_2 by the enzyme RuBisCo into 3-phosphoglycerate. However, due to the dual carboxylase/oxygenase activity of RuBisCo, an amount of the substrate is oxidized rather than carboxylated resulting in loss of substrate and consumption of energy, in what is known as photorespiration. In order to bypass the photorespiration pathway, C_4 plants have developed a mechanism to efficiently deliver CO_2 to the RuBisCO enzyme. They utilize their specific leaf anatomy where chloroplasts exist not only in the mesophyll cells in the outer part of their leaves but in the bundle sheath cells as well. Instead of direct fixation in the Calvin cycle, CO_2 is converted to a 4-carbon organic acid which has the ability to regenerate CO_2 in the chloroplasts of the bundle sheath cells. Bundle sheath cells can then utilize this CO_2 to generate carbohydrates by the conventional C_3 pathway.

The first step in the pathway is the conversion of pyruvate to PEP by the enzyme pyruvate-phosphate dikinase (pyruvate, orthophosphate dikinase); this reaction requires inorganic phosphate and ATP plus pyruvate, giving phosphoenolpyruvate, AMP, and PPi (inorganic pyrophosphate) as products. The next step is the fixation of CO_2 by the enzyme phosphoenolpyruvate carboxylase. Both of these steps occur in the mesophyll cells:

pyruvate + Pi + ATP ? PEP + AMP + PPi

PEP carboxylase + PEP + CO_2 ? oxaloacetate

PEP carboxylase has a lower Km for CO_2—and hence higher affinity—than Rubisco. Furthermore, O_2 is a very poor substrate for this enzyme. Thus, at relatively low concentrations of CO_2, most CO_2 will be fixed by this pathway.

The product is usually converted to malate, a simple organic compound that is transported to the bundle-sheath cells surrounding a nearby vein, where it is decarboxylated to release CO_2, which enters Calvin cycle. The decarboxylation leaves pyruvate, which is transported back to the mesophyll cell.

Since every CO_2 molecule has to be fixed twice, the C_4 pathway is more energy-consuming than the C_3 pathway. The C_3 pathway requires 18 ATP for the synthesis of one molecule of glucose while the C_4 pathway requires 30 ATP. But since otherwise tropical plants lose more than half of photosynthetic carbon in photorespiration, the C_4 pathway is an adaptive mechanism for minimizing the loss.

There are several variants of this pathway:

The 4-carbon acid transported from mesophyll cells may be malate as above, or may be aspartate.

The 3-carbon acid transported back from bundle-sheath cells may be pyruvate as above, or alanine.

The enzyme which catalyses decarboxylation in bundle-sheath cells differs. In maize and sugarcane, the enzyme is NADP-malic enzyme, in millet, it is NAD-malic enzyme, and in *Panicum maximum* it is PEP carboxykinase.

C_4 LEAF ANATOMY

The C_4 plants possess a characteristic leaf anatomy. Their vascular bundles are surrounded by two rings of cells. The inner ring, called Bundle Sheath Cells, contain starch-rich chloroplasts lacking grana which differ from those in mesophyll cells present as the outer ring. Hence, the chloroplasts are called dimorphic. This peculiar anatomy is called Kranz Anatomy (Kranz-Crown/Halo). The primary function of the Kranz is to provide a site in which carbon dioxide can be concentrated around RuBisCO, thus reducing photorespiration. In order to facilitate the maintenance of a significantly higher carbon dioxide concentration in the bundle sheath compared to the mesophyll, the boundary layer of the Kranz has a low conductance to carbon dioxide, a property which may be enhanced by the presence of suberin.

Although most C_4 plants exhibit Kranz anatomy, there are a number of species which operate a limited C_4 cycle without any distinct bundle sheath tissue. *Suaeda aralocaspica* (formerly known as *Borszczowia aralocaspica*), *Bienertia cycloptera* and *Bienertia sinuspersici* (all chenopods) are terrestrial plants which inhabit dry, salty depressions in the deserts of south-east Asia. These plants have been shown to operate single-cell C_4 carbon dioxide concentrating mechanisms which are unique amongst the known C_4 mechanisms. Although the cytology of both species differ slightly, the basic principle is that fluid filled vacuoles are employed to divide the cell into to separate areas. Carboxylation enzymes in the cytosol can therefore be kept separate from decarboxylase enzymes and RuBisCo in the chloroplasts, and a diffusive barrier can be established between the chloroplasts (which contain RuBisCO) and the cytosol. This enables a bundle-sheath type area and a mesophyll type area to be established within a single cell. Although this does allow a limited C_3 cycle to operate, it is relatively inefficient, with much leakage of CO_2 from around RuBisCO occurring. There is also evidence for the non-Kranz aquatic macrophyte Hydrilla verticillata exhibiting inducible C_4 photosynthesis under warm conditions, although the mechanism by which CO_2 leakage from around RuBisCO is minimised is currently uncertain.

THE EVOLUTION AND ADVANTAGES OF THE C_4 PATHWAY

C_4 plants have a competitive advantage over plants possessing the more common C_3 carbon fixation pathway under conditions of drought, high temperatures and nitrogen or carbon dioxide limitation. 97% of the water taken up by C_3 plants is lost through transpiration compared to a much lower proportion in C_4 plants, demonstrating their advantage in a dry environment.

C_4 carbon fixation has evolved on up to 40 independent occasions in different groups of plants, making it an example of convergent evolution Plants which use C_4 metabolism include sugarcane, maize, sorghum, finger millet, amaranth, and switchgrass. C_4 plants arose around 25 to 32 million years ago during the Oligocene (precisely when is difficult to determine) and did not become ecologically significant until around 6 to 7 million years ago, in the Miocene Period. Today they represent about 5% of Earth's plant biomass and 1% of its known plant species However, they account for around 30% of terrestrial carbon fixation These species are concentrated in the tropics (below latitudes of 45°) where the high air temperature contributes to higher possible levels of oxygenase activity by RuBisCO, which increases rates of photorespiration in C_3 plants.

15

Phycobilins

INTRODUCTION

Phycobilins are complex photoreceptor pigments – open-chain tetrapyrroles that are structurally related to mammalian bile pigments. Phytochromes are phycobilin-protein pigments involved in floral induction. There are two classes of phycobilins and they occur only in Cyanobacteria and Rhodophyta. The phycobilin component is similar to the porphyrins without a metallic atom. Water-soluble phycobilin pigments are found in the stroma of the chloroplast. In at least two groups of algae, phycobiliproteins are aggregated in a highly ordered protein complex called a phycobilisome (PBS), making these phycobilins unique among photosynthetic pigments.

Phycobilisomes are attached to the cytosol (stromal) face of the thylakoid. Extending into the cytosol, the phycobilisomes consist of a cluster of phycobilin pigments including phycocyanin (blue) and phycoerythrin (red) attached by their phycobiliproteins. These particles serve as light-energy antennae for photosynthesis. Phycobilisomes preferentially funnel light energy into photosystem II for the splitting of water and generation of oxygen. While many photosynthetic eubacteria possess photosystem I to oxidize reduced molecules such as H2S, only Cyanobacteria have photosystem II. The evolution of photosystem II apparently occured in Cyanobacteria.

The bluish pigment phycocyanin is found in Cyanobacteria, giving them their misleading common name of "blue-green algae". Different species of cyanobacteria possess differing ratios of phyocyanin and phycoerythrin. Cyanobacteria such as *Hammatoidea, Heterohormogonium, Albrightia, Scytonematopsis, Thalopophila, Myxocarcina* and *Colteronema* confer colours from red to purple on thermal springs and geyser pools. The ratio of phycocyanin and phycoerythrin can be environmentally altered. Cyanobacteria which are raised in green light typically develop more phycoerythrin and become red. The same Cyanobacteria grown in red light become bluish-green. This reciprocal colour change has been named 'chromatic adaptation'.

Phycoerythrin is an accessory photoreceptor pigment found in the Rhodophyta ("red algae"). Phycoerythrin is associated with chlorophyll in the Rhodophyta, and enables them to be photosynthetically efficient in deep water where blue light predominates. The longer wavelength red portion of the spectrum that activate green chlorophyll pigments do not penetrate the deeper water of the photic zone, so green algae cannot survive at depth where red algae thrive.

Three major classes of photosynthetic pigments occur among the algae: chlorophylls, carotenoids (carotenes and xanthophylls) and phycobilins. The phycobilins and the carotenoid peridinin are water soluble. In the Cryptophyta, the phycobilin pigments are found in the spaces between the thylakoids, not in phycobilisomes as they are in the Cyanobacteria and Rhodophyta. Alpha-carotene, and the xanthophyll, diatoxanthan, combine with the proteinaceous phycobilin pigments phycoerythrin and phycocyanin. Chlorophyll-a and chlorophyll-c1 are the main photosynthetic pigments of the Cryptophyta, and chlorophyll-b is never present.

The photosynthesizing ability of eukaryotes was made possible by one or more endosymbiotic associations between heterotrophic eukaryotes and photosynthetic prokaryotes (or their descendents). Several primary endosymbioses occurred between eukaryotes and blue green algae. In one of the lineages, the photosynthetic organism lost much of its genetic independence and became functionally and genetically integrated as plastids – chloroplasts within the host cell. At least two types of protists – chloroarachniophytes and cryptomonads-acquired 'plastids' by forming symbioses with eukaryotic algae. Such acquisitions are referred to as secondary symbioses. Palintomic: a palintomic cell division is a cell division without a preceding compensatory increase in size.

Parasitophorous vacuole: a vacuole within which intracellular parasites grow and develop.

Pellicle: a generic term referring to the peripheral structures which maintain cell shape and integrity. The pellicle typically includes the plasma membrane, an alveolar layer, fibrous meshes and microtubules.

Pelta: in Metamonada, an anterior wall of microtubules, typically covering the nucleus. There is apparently a closely spaced series of bridges across the membrane which connect the microtubules of the axostyle and pelta. In other Protists, the term is used for any network of microtubules which forms part of the pellicle.

Peroxisome: The peroxisome is a single-membrane organelle present in nearly all eukaryotic cells. One of the most important metabolic processes of the peroxisome is the b-oxidation of long and very long chain fatty acids. The peroxisome is also involved in bile acid synthesis, cholesterol synthesis, plasmalogen synthesis, amino acid metabolism, and purine metabolism. The peroxisome for the scientist. Peroxisomes are formed by self-assembly and are not budded off from the Golgi (like lysosomes) or the endoplasmic reticulum. Peroxisomes contain oxidative enzymes, such as D-amino acid oxidase, urate oxidase, and catalase. Peroxisomes are distinguished by a crystalline structure inside a sac which also contains amorphous gray material. They are self replicating, like the mitochondria. Peroxisomes frequently function to detoxify the cell by eliminating substances like hydrogen peroxide, or other metabolites. Peroxisomes have membrane proteins that are critical for peroxisomal function, to import proteins into their interiors, proliferate or segregate to daughter cells.

Phagocytosis: a method of food ingestion in which a food particle is encapsulated in a membranous food vacuole as it passes through the plasma membrane. The food vacuole then fuses with intracellular vacuoles containing digestive enzymes.

Phycobilin: a class of light-sensitive ligands bound to phycobiliproteins and used as accessory light-gathering pigments in algae. The mechanism is described in the entries for *phycobiliprotein* and *phycobilisome*. The phycobilins are, like chlorophyll, tetrapyrrole structures. However, unlike chlorophyll, the pyrrole rings are laid out linearly. The detailed structure of four phycobilins commonly found in algae are shown in the figure. The phycobilins have light absorption maxima which vary considerably depending on their exact chemical environment. It is also wrong to assume that any particular phycobilin species is exclusively bound to the phycobiliprotein with a similar name. There is considerable variation.

Phycobiliprotein: "water soluble fluorescent proteins derived from cyanobacteria and eukaryotic algae. In these organisms, they are used as accessory or antenna pigments for photosynthetic light collection. They absorb energy in portions of the visible spectrum that are poorly utilized by chlorophyll and, through fluorescence energy transfer, convey the energy to chlorophyll at the photosynthetic reaction center. The phycobiliproteins are composed of a number of subunits, each having a protein backbone to which linear tetrapyrrole chromophores are covalently bound. All phycobiliproteins contain either phycocyanobilin or phycoerythrobilin chromophores, and may also contain one of three minor bilins; phycourobilin, cryptoviolin [= phycobiliviolin?], or the 697-nm bilin. Each bilin has unique spectral characteristics, which may be further modified by interactions of the subunits and of the chromophore with the apoprotein. The phycobiliproteins in many algae are arranged in subcellular structures called phycobilisomes. These structures allow the pigments to be arranged geometrically in a manner which helps to optimize the capture of light and transfer of energy.

Phycobilisome: a structure composed of several phycobiliprotein complexes attached to the outer surface of the thylakoid membranes in some cyanobacteria and in the chloroplasts of glaucophytes and red algae. The phycobilisome is so arranged that a very broad range of light frequencies can be modulated by the phycobilin pigments and the energy transferred to chlorophyll in the light reactions of photosystem II (located inside the thylakoid membrane).

Piriform: pear-shaped. Often spelled *pyriform.*

Plaque, centrosomal: A structure present in organisms, especially Fungi, which have a spindle pole body, rather than a conventional centrosome. The plaque is made up of three distinct sections: outer, inner and central. The central plaque is anchored in the plane of the nuclear envelope, the outer plaque nucleates the cytoplasmic microtubules, and the inner plaque nucleates the spindle microtubules in the nucleus.

Plasma membrane: the principal outer membrane of the cell which encloses the cytoplasm.

Plasmalemma: same as plasma membrane.

Polar filament: in the Microsporidia, a long, coiled series of filaments which extend the polar tube to infect host cells.

Polar ring: ring-shaped microtubular structures in the apical complexes of Apicomplexa. The polar rings, like most other things in the apical complex, are activated by the release of sequestered calcium ions into the parasite's cytoplasm. The presumably act in host cell invasion. The relative movement of polar rings in one system is illustrated at the glossary entry on the conoid. *See also* entry at subpellicular microtubules. As shown in those images, the subpellicular microtubules (if present) extend like lines of longitude from regularly spaced attachment points on the outer polar ring.

Polar tube: in the Microsporidia, the equivalent of a hypodermic needle. This organelle is inserted into a host cell and infectious sporoplasm is injected into the host.

Polaroplast: in the Microsporidia, a complex structure of layered vesicles or membranes associated with the base of the polar tube.

Preaxostyle: a helmet-shaped microtubular body which caps the anterior pole of the nucleus.

Pronucleus: the nucleus of a gamete, after fertilization but before complete fusion of the two haploid nuclei.

Pseudopodium: a temporary protrusion on the cell surface. Initially the cell extends a membrane process known as a lamellopodium. This is accompanied by controlled polymerization of actin filaments at the leading edge and the subsequent incorporation of these actin filaments into bundles and networks. The origin of the actual force that propels the cell forward is unknown, although it is thought to be the polymerization of the actin filaments.

RADIOLARIA

Radiolaria: a taxonomic grouping approximating the Radiolaria of Haeckel (1887). It includes the Acantharea, Phaeodarea, and Polycystina, all of which are united by numerous characters such as a capsule which divides the cytoplasm into intra and extracapsular spaces, a mineralized cytoskeleton with externally-projecting spicules, numerous filipodia, and perhaps similarities of life cycle.

rDNA: the DNA which codes for ribosomal RNA. If that term isn't familiar, see the Cell Biology summary at Eubacteria. However, there are some significant differences. In contrast to prokaryotes, the four RNAs contained in eukaryotic ribosomes are coded by two types of genetic units which are generally not linked. Each type occurs in its own randomly repeated clusters. The larger unit, the rDNA, is transcribed by RNA polymerase I as a single precursor containing the small subunit (18S) rRNA, 5.8S rRNA and the large subunit (28S) rRNA, each bracketed with spacer sequences. The second type of unit codes for 5S rRNA and is transcribed by RNA polymerase III.

recurrent: oriented opposite the direction of motion.

Reticulopodia: see reticulose.

Reticulose: Forming a network; characterized by a reticulated structure. Reticulose pseudopods are pseudopods in which the individual pseudopodia blend together and form irregular meshes.

Rhoptry: The rhoptries are club-shaped secretory organelles, often located near the apical end of Apicomplexan intracellular parasites. Rhoptries are secreted during host cell invasion, and rhoptry proteins are found within the lumen and the membrane of very early stages of the forming *parasitophorous vacuole.*

Ribosomal RNA: ribosomes are the small, but incredibly complex nucleoprotein complexes responsible for protein synthesis. They bind to mRNA molecules from the nucleus and physically move along the molecule, "reading" the code on the mRNA and attaching amino acids to the growing peptide (protein) chain. In eukaryotes there are three, quite distinctive RNA species bound up in the ribosome. These are known by their "Svedberg" numbers, a possibly obsolete measure of relative movement in centrifugation through a density gradient. The three species are the 5S, 18S and 28S RNAs, *vide infra*. Mitochondria produce their own ribosomal RNAs, the 16S and 23S rRNAs. The foolishness of using mitochondrial rRNA for phylogenetic purposes is addressed at length elsewhere.

Ribosomal RNA, 18S or ssu RNA: the RNA molecule associated with the small ribosomal subunit. The secondary structure of typical 18S rRNA. The latter are generally referred to by the standard nomenclature shown in the figure, *vis., V1, V2,* etc. The other numbers in the figure refer to stems, but do not appear to match the standard nomenclature for stems used by many other workers.

Ribosome: the cellular organelle responsible for translating mRNA into protein. Eukaryotic ribosomes are complexes of specialized RNA species and numerous proteins.

RNA polymerase: any of the enzyme complexes directly responsible for transcription — the manufacture of RNA from the DNA template. RNA polymerase I is specialized for the synthesis of rRNA. RNA polymerase II is used to synthesize mRNAs or their precursors. RNA polymerase III is used to transcribe a single species, the 5S RNA of the large ribosomal subunit.

Rostellum: some commensal oxymonads (Oxymonadidae) have an elongate anterior structure which terminates in a holdfast, through which the cells attach to the gut wall. This is referred to as the rostellum. *See* image at Oxymonadidae.

RNA: ribosomal RNA, *q.v.*

Rubisco: an acronym for ribulose bisphosphate carboxylase/ oxygenase. Photosynthesis is the process of fixing atmospheric carbon dioxide and transforming it into organic carbon. Rubisco is the enzyme which actually does the trick. Specifically, rubisco attaches CO_2 to ribulose bisphosphate, a five carbon sugar. It then splits the molecule into two 3-carbon phosphoglycerates which feed into a number of different metabolic pathways. Rubisco is unusually slow and inefficient. It fixes only about three carbon dioxide molecules per second, compared to 1000+ for an average metabolic enzyme. It is also easily confused by other substrates, notably oxygen, and makes a remarkable number of errors. Perhaps the evolved design, as bad as it is, can be no better. Rather than improving the process, plants simply make enormous quantities of enzyme. Rubisco is, in fact, the most common protein on earth. As much as 50% of the mass of each chloroplast is rubisco. Plants and algae build rubisco in compact octamers, with each monomer containing two peptides. The active site contains a magnesium ion bound by three amino acids. One of these is a uniquely modified lysine with an extra carboxyl group added to the end of its side chain. In plant cells, this activator group, is attached to rubisco during the day, turning the enzyme "on," and removed at night, turning the enzyme "off." The exposed side of the magnesium ion binds to both ribulose bisphosphate and the substrate carbon dioxide molecule.

Visual Phototransduction

Visual phototransduction is a process by which light is converted into electrical signals in the rod cells, cone cells and photosensitive ganglion cells of the retina of the eye.

The visual cycle is the biological conversion of a photon into an electrical signal in the retina. This process occurs via G-protein coupled receptors called opsins which contain the chromophore 11-cis retinal. 11-cis retinal is covalently linked to the opsin receptor via a Schiff base forming a retinylidene protein.

When struck by a photon, 11-cis retinal undergoes photoisomerization to all-trans retinal which changes the conformation of the opsin GPCR leading to signal transduction cascades which causes closure of a cyclic GMP-gated cation channel, and hyperpolarization of the photoreceptor cell.

Following isomerization and release from the opsin protein, all-trans retinal is reduced to all-trans retinol and travels back to the retinal pigment epithelium to be "recharged". It is first esterified by lecithin-retinol acyltransferase (LRAT) and then converted to 11-cis retinol by the isomerohydrolase RPE65. The isomerase activity of RPE65 has been shown, but it is still uncertain whether it also acts as a hydrolase. Finally, it is oxidized to 11-cis retinal before traveling back to the rod outer segment where it can again be conjugated to an opsin to form a new, functional visual pigment (rhodopsin).

PHOTORECEPTORS

The process of phototransduction is a complicated one, and in order to understand it, one must have an understanding of the structure of the photoreceptor cells involved in vision: the rods and cones. These cells contain a chromophore (11-*cis*-retinal, the aldehyde of Vitamin A1 and light-absorbing portion) bound to a cell membrane protein, opsin. Rods deal with low light level and do not mediate colour vision. Cones, on the other hand, can code the colour of an image through comparison of the outputs of the three different types of cones. Each cone type responds best to certain wavelengths, or colours, of light because each type has a slightly different opsin. The three types of cones are L-cones, M-cones and S-cones that respond optimally to long wavelengths (reddish colour), medium wavelengths (greenish colour), and short wavelengths (bluish colour) respectively.

Process

To understand the photoreceptor's behaviour to light intensities, it is necessary to understand the roles of different currents.

There is an ongoing outward potassium current through nongated K^{+-} selective channels. This outward current tends to hyperpolarize the photoreceptor at around -70 mV (the equilibrium potential for K^+).

There is also an inward sodium current carried by cGMP-gated sodium channels. This so-called 'dark current' depolarizes the cell to around -40 mV. Note that this is significantly more depolarized than most other neurons.

A high density of Na^+ - K^+ pumps enables the photoreceptor to maintain a steady intracellular concentration of Na^+ and K^+.

In the Dark

Photoreceptor cells are strange cells because they are depolarized in the dark, i.e. light hyperpolarizes and switches off these cells, and it is this 'switching off' that activates the next cell and sends an excitatory signal down the neural pathway.

In the dark, cGMP levels are high and keep cGMP-gated sodium channels open allowing a steady inward current, called the dark current. This dark current keeps the cell depolarised at about -40 mV.

The depolarization of the cell membrane opens voltage-gated calcium channels. An increased intracellular concentration of Ca^{2+} causes vesicles containing special chemicals, called neurotransmitters, to merge with the cell membrane, therefore releasing the neurotransmitter into the synaptic cleft, an area between the end of one cell and the beginning of another neuron. The neurotransmitter released is glutamate, an excitatory neurotransmitter.

In the cone pathway glutamate:

- Hyperpolarizes on-center bipolar cells. Glutamate that is released from the photoreceptors in the dark binds to metabotropic glutamate receptors (mGluR6), which, through a G-protein coupling mechanism, causes non-specific cation channels in the cells to close, thus hyperpolarizing the bipolar cell.
- Depolarizes off-center bipolar cells. Binding of glutamate to ionotropic glutamate receptors results in an inward cation current that depolarizes the bipolar cell.

In the Light

Representation of molecular steps in photoactivation (modified from Leskov et al., 2000). Depicted is an outer membrane disk in a rod. Step 1: Incident photon (h?) is absorbed and activates a rhodopsin by conformational change in the disk membrane to R*. Step 2: Next, R* makes repeated contacts with transducin molecules, catalyzing its activation to G* by the release of bound GDP in exchange for cytoplasmic GTP. The a and ? subunits Step 3: G* binds inhibitory ? subunits of the phosphodiesterase (PDE) activating its a and ß subunits. Step 4: Activated PDE hydrolyzes cGMP. Step 5: Guanylyl cyclase (GC) synthesizes cGMP, the second messenger in the phototransduction cascade. Reduced levels of cytosolic cGMP cause cyclic nucleotide gated channels to close preventing further influx of Na^+ and $Ca2^+$.

DEACTIVATION OF THE PHOTOTRANSDUCTION CASCADE

GTPase Activating Protein (GAP) interacts with the alpha subunit of transducin, and causes it to hydrolyse its bound GTP to GDP, and thus halts the action of phosphodiesterase, stopping the transformation of cGMP to GMP.

Guanylate Cyclase Activating Protein (GCAP) is a calcium binding protein, and as the calcium levels in the cell have decreased, GCAP dissociates from its bound calcium ions, and interacts with Guanylate Cyclase, activating it. Guanylate Cyclase then proceeds to transform GTP to cGMP, replenishing the cell's cGMP levels and thus reopening the sodium channels that were closed during phototransduction.

Finally, Metarhodopsin II is deactivated. Recoverin, another calcium binding protein, is normally bound to Rhodopsin Kinase when calcium is present. When the calcium levels fall during phototransduction, the calcium dissociates from recoverin, and rhodopsin kinase is released, when it proceeds to phosphorylate metarhodopsin II, which decreases its affinity for transducin. Finally, arrestin, another protein, binds the phosphorylated metarhodopsin II, completely deactivating it. Thus, finally, phototransduction is deactivated, and the dark current and glutamate release is restored. It is this pathway, where Metarhodopsin II is phosphorylated and bound to arrestin and thus deactivated, which is thought to be responsible for the S2 component of dark adaptation. The S2 component represents a linear section of the dark adaptation function present at the beginning of dark adaptation for all bleaching intensities.

All-*trans* retinal is transported to the pigment epithelial cells to be reduced to all-*trans* retinol, the precursor to 11-*cis* retinal. This is then transported back to the rods. All-*trans* retinol cannot be synthesised by humans and must be supplied by vitamin A in the diet. Deficiency of all-*trans* retinol can lead to night blindness. This is part of the bleach and recycle process of retinoids in the photoreceptors and retinal pigment epithelium.

A pigment is a material that changes the colour of light it reflects as the result of selective colour absorption. This physical process differs from fluorescence, phosphorescence, and other forms of luminescence, in which the material itself emits light.

Many materials selectively absorb certain wavelengths of light. Materials that humans have chosen and developed for use as pigments usually have special properties that make them ideal for colouring other materials. A pigment must have a high tinting strength relative to the materials it colours. It must be stable in solid form at ambient temperatures.

For industrial applications, as well as in the arts, permanence and stability are desirable properties. Pigments that are not permanent are called fugitive. Fugitive pigments fade over time, or with exposure to light, while some eventually blacken.

Pigments are used for colouring paint, ink, plastic, fabric, cosmetics, food and other materials. Most pigments used in manufacturing and the visual arts are dry colourants, usually ground into a fine powder. This powder is added to a vehicle (or matrix), a relatively neutral or colourless material that acts as a binder.

A distinction is usually made between a pigment, which is insoluble in the vehicle (resulting in a suspension), and a dye, which either is itself a liquid or is soluble in its vehicle (resulting in a solution). A colourant can be both a pigment and a dye depending on the vehicle it is used in. In some cases, a pigment can be manufactured from a dye by precipitating a soluble dye with a metallic salt. The resulting pigment is called a lake pigment.

PHYSICAL BASIS

A wide variety of wavelengths (colours) encounter a pigment. This pigment absorbs red and green light, but reflects blue, creating the colour blue.

Pigments appear the colours they are because they selectively reflect and absorb certain wavelengths of light. White light is a roughly equal mixture of the entire visible spectrum of light. When this light encounters a pigment, some wavelengths are absorbed by the chemical bonds and substituents of the pigment, and others are reflected. This new reflected light spectrum creates the appearance of a colour. Ultramarine reflects blue light, and absorbs other colours. Pigments, unlike fluorescent or phosphorescent substances, can only subtract wavelengths from the source light, never add new ones.

The appearance of pigments is intimately connected to the colour of the source light. Sunlight has a high colour temperature, and a fairly uniform spectrum, and is considered a standard for white light. Artificial light sources tend to have great peaks in some parts of their spectrum, and deep valleys in others. Viewed under these conditions, pigments will appear different colours.

Sunlight encounters Rosco R80 "Primary Blue" pigment. The product of the source spectrum and the reflectance spectrum of the pigment results in the final spectrum, and the appearance of blue.

Colour spaces used to represent colours numerically must specify their light source. Lab colour measurements, unless otherwise noted, assume that the measurement was taken under a D65 light source, or "Daylight 6500 K", which is roughly the colour temperature of sunlight.

Other properties of a colour, such as its saturation or lightness, may be determined by the other substances that accompany pigments. Binders and fillers added to pure pigment chemicals also have their own reflection and absorption patterns, which can affect the final spectrum. Likewise, in pigment/binder mixtures, individual rays of light may not encounter pigment molecules, and may be reflected as is. These stray rays of source light contribute to the saturation of the colour. Pure pigment allows very little white light to escape, producing a highly saturated colour. A small quantity of pigment mixed with a lot of white binder, however, will appear desaturated and pale, due to the high quantity of escaping white light.

Pigment Groups

- *Arsenic pigments:* Paris Green
- *Carbon pigments:* Carbon Black, Ivory Black, Vine Black, Lamp Black
- *Cadmium pigments:* cadmium pigments, Cadmium Green, Cadmium Red, Cadmium Yellow, Cadmium Orange.
- *Iron oxide pigments:* Caput Mortuum, oxide red, Red Ochre, Sanguine, Venetian Red.
- Prussian blue
- *Chromium pigments:* Chrome Green, Chrome Yellow
- *Cobalt pigments:* Cobalt Blue, Cerulean Blue, Cobalt Violet, Aureolin
- *Lead pigments:* lead white, Naples yellow, Cremnitz White, red lead
- *Copper pigments:* Paris Green, Verdigris, Viridian, Egyptian Blue, Han Purple
- *Titanium pigments:* Titanium White, Titanium Beige, Titanium yellow, Titanium Black
- *Ultramarine pigments:* Ultramarine, Ultramarine Green Shade, French Ultramarine
- *Mercury pigments:* Vermilion
- *Zinc pigments:* Zinc White

- *Clay earth pigments (which are also iron oxides):* Raw Sienna, Burnt Sienna, Raw Umber, Burnt Umber, Yellow Ochre.
- *Biological origins:* Alizarin, Alizarin Crimson, Gamboge, Indigo, Indian Yellow, Cochineal Red, Tyrian Purple, Rose madder.
- *Other Organic:* Pigment Red 170, Phthalo Green, Phthalo Blue, Quinacridone Magenta.

BIOLOGICAL PIGMENTS

The monarch butterfly's distinctive pigmentation reminds potential predators that it is poisonous.

In biology, a pigment is any coloured material of plant or animal cells. Many biological structures, such as skin, eyes, fur and hair contain pigments (such as melanin) in specialized cells called chromatophores. Many conditions affect the levels or nature of pigments in plant, animal, some protista, or fungus cells. For instance, Albinism is a disorder affecting the level of melanin production in animals.

Pigment colour differs from structural colour in that it is the same for all viewing angles, whereas structural colour is the result of selective reflection or iridescence, usually because of multilayer structures. For example, butterfly wings typically contain structural colour, although many butterflies have cells that contain pigment as well.

History

Naturally occurring pigments such as ochres and iron oxides have been used as colourants since prehistoric times. Archaeologists have uncovered evidence that early humans used paint for aesthetic purposes such as body decoration. Pigments and paint grinding equipment believed to be between 350,000 and 400,000 years old have been reported in a cave at Twin Rivers, near Lusaka, Zambia.

Before the Industrial Revolution, the range of colour available for art and decorative uses was technically limited. Most of the pigments in use were earth and mineral pigments, or pigments of biological origin. Pigments from unusual sources such as botanical materials, animal waste, insects, and mollusks were harvested and traded over long distances. Some colours were costly or impossible to mix with the range of pigments that were available. Blue and purple came to be associated with royalty because of their expense.

Biological pigments were often difficult to acquire, and the details of their production were kept secret by the manufacturers. Tyrian Purple is a pigment made from the mucus of one of several species of Murex snail. Production of Tyrian Purple for use as a fabric dye began as early as 1200 BCE by the Phoenicians, and was continued by the Greeks and Romans until 1453 CE, with the fall of Constantinople The pigment was expensive and complex to produce, and items coloured with it became associated with power and wealth. Greek historian Theopompus, writing in the 4th century BCE, reported that "purple for dyes fetched its weight in silver at Colophon [in Asia Minor].

Mineral pigments were also traded over long distances. The only way to achieve a deep rich blue was by using a semi-precious stone, lapis lazuli, to produce a pigment known as ultramarine, and the best sources of lapis were remote. Flemish painter Jan Van Eyck, working in the 15th century, did not ordinarily include blue in his paintings. To have one's portrait commissioned and painted with ultramarine blue was considered a great luxury. If a patron wanted blue, they were forced to pay extra. When Van Eyck used lapis, he never blended it with other colours. Instead he applied it in pure form, almost as a decorative glaze. The prohibitive price of lapis lazuli forced artists to seek less expensive replacement pigments, both mineral (azurite, smalt) and biological (indigo).

Spain's conquest of a New World empire in the 16th century introduced new pigments and colours to peoples on both sides of the Atlantic. Carmine, a dye and pigment derived from a parasitic insect found in Central and South America, attained great status and value in Europe. Produced from harvested, dried, and crushed cochineal insects, carmine could be used in fabric dye, body paint, or in its solid lake form, almost any kind of paint or cosmetic.

Natives of Peru had been producing cochineal dyes for textiles since at least 700 CE but Europeans had never seen the colour before. When the Spanish invaded the Aztec empire in what is now Mexico, they were quick to exploit the colour for new trade opportunities. Carmine became the region's second most valuable export next to silver. Pigments produced from the cochineal insect gave the Catholic cardinals their vibrant robes and the English "Redcoats" their distinctive uniforms. The true source of the pigment, an insect, was kept secret until the 18th century, when biologists discovered the source.

While Carmine was popular in Europe, blue remained an exclusive colour, associated with wealth and status. The 17th century Dutch master Johannes Vermeer often made lavish use of lapis lazuli, along with Carmine and Indian Yellow, in his vibrant paintings.

Development of Synthetic Pigments

The earliest known pigments were natural minerals. Natural iron oxides give a range of colours and are found in many Paleolithic and Neolithic cave paintings. Two examples include Red Ochre, anhydrous Fe_2O_3, and the hydrated Yellow Ochre ($Fe_2O_3.H_2O$) Charcoal, or carbon black, has also been used as a black pigment since prehistoric times.

Two of the first synthetic pigments were white lead (basic lead carbonate, $(PbCO_3)_2Pb(OH)_2$) and blue frit (Egyptian Blue). White lead is made by combining lead with vinegar (acetic acid, CH_3COOH) in the presence of CO_2. Blue frit is calcium copper silicate and was made from glass coloured with a copper ore, such as malachite. These pigments were used as early as the second millennium BCE.

The Industrial and Scientific Revolutions brought a huge expansion in the range of synthetic pigments, pigments that are manufactured or refined from naturally occurring materials, available both for manufacturing and artistic expression. Because of the expense of Lapis Lazuli, much effort went into finding a less costly blue pigment.

Prussian Blue was the first modern synthetic pigment, discovered by accident in 1704. By the early 19th century, synthetic and metallic blue pigments had been added to the range of blues, including French ultramarine, a synthetic form of lapis lazuli, and the various forms of Cobalt and Cerulean Blue. In the early 20th century, organic chemistry added Phthalo Blue, a synthetic, organic pigment with overwhelming tinting power.

Discoveries in colour science created new industries and drove changes in fashion and taste. The discovery in 1856 of mauveine, the first aniline dye, was a forerunner for the development of hundreds of synthetic dyes and pigments. Mauveine was discovered by an 18-year-old chemist named William Henry Perkin, who went on to exploit his discovery in industry and become wealthy. His success attracted a generation of followers, as young scientists went into organic chemistry to pursue riches. Within a few years, chemists had synthesized a substitute for madder in the production of Alizarin Crimson. By the closing decades of the 19th century, textiles, paints, and other commodities in colours such as red, crimson, blue, and purple had become affordable.

Development of chemical pigments and dyes helped bring new industrial prosperity to Germany and other countries in northern Europe, but it brought dissolution and decline elsewhere. In Spain's former New World empire, the production of cochineal colours employed thousands of low-paid workers. The Spanish monopoly on cochineal production had been worth a fortune until the early 1800s, when the Mexican War of Independence and other market changes disrupted production Organic chemistry delivered the final blow for the cochineal colour industry. When chemists created inexpensive substitutes for carmine, an industry and a way of life went into steep decline

New sources for Historic Pigments

Before the Industrial Revolution, many pigments were known by the location where they were produced. Pigments based on minerals and clays often bore the name of the city or region where they were mined. Raw Sienna and Burnt Sienna came from Siena, Italy, while Raw Umber and Burnt Umber came from Umbria. These pigments were among the easiest to synthesize, and chemists created modern colours based on the originals that were more consistent than colours mined from the original ore bodies. But the place names remained.

Historically and culturally, many famous natural pigments have been replaced with synthetic pigments, while retaining historic names. In some cases the original colour name has shifted in meaning, as a historic name has been applied to a popular modern colour. By convention, a contemporary mixture of pigments that replaces a historical pigment is indicated by calling the resulting colour a hue, but manufacturers are not always careful in maintaining this distinction. The following examples illustrate the shifting nature of historic pigment names:

Titian used the historic pigment Vermilion to create the reds in the great fresco of Assunta, completed c. 1518.

- Indian Yellow was once produced by collecting the urine of cattle that had been fed only mango leaves. Dutch and Flemish painters of the 17th and 18th centuries favored it for its luminescent qualities, and often used it to represent sunlight. In *Girl with a Pearl Earring*, Vermeer's patron remarks that Vermeer used "cow piss" to paint his wife. Since mango leaves are nutritionally inadequate for cattle, the practice of harvesting Indian Yellow was eventually declared to be inhumane. Modern Indian Yellow Hue is a mixture of synthetic pigments.
- Ultramarine, originally the semi-precious stone lapis lazuli, has been replaced by an inexpensive modern synthetic pigment manufactured from aluminium silicate with sulfur impurities. At the same time, Royal Blue, another name once given to tints produced from lapis lazuli, has evolved to signify a much lighter and brighter colour, and is usually mixed from Phthalo Blue and titanium dioxide, or from inexpensive synthetic blue dyes. Since synthetic ultramarine is chemically identical with lapis lazuli, the "hue" designation is not used. French Blue, yet another historic

name for ultramarine, was adopted by the textile and apparel industry as a colour name in the 1990s, and was applied to a shade of blue that has nothing in common with the historic pigment French ultramarine.

- Vermilion, a toxic mercury compound favored for its deep red-orange colour by old master painters such as Titian, has been replaced by convenient mixtures of synthetic, inorganic pigments. Although genuine Vermilion paint can still be purchased for fine arts and art conservation applications, few manufacturers make it, because of legal liability issues. Few artists buy it, because it has been superseded by modern pigments that are both less expensive and less toxic, as well as less reactive with other pigments. As a result, genuine Vermilion is almost unavailable. Modern vermilion colours are properly designated as Vermilion Hue to distinguish them from genuine Vermilion.

Manufacturing and Industrial Standards

Before the development of synthetic pigments, and the refinement of techniques for extracting mineral pigments, batches of colour were often inconsistent. With the development of a modern colour industry, manufacturers and professionals have cooperated to create international standards for identifying, producing, measuring, and testing colours.

First published in 1905, the Munsell Colour System became the foundation for a series of colour models, providing objective methods for the measurement of colour. The Munsell system describes a colour in three dimensions, hue, value (lightness), and chroma (colour purity), where chroma is the difference from gray at a given hue and value.

By the middle years of the 20th century, standardized methods for pigment chemistry were available, part of an international movement to create such standards in industry. The International Organization for Standardization (ISO) develops technical standards for the manufacture of pigments and dyes. ISO standards define various industrial and chemical properties, and how to test for them. The principal ISO standards that relate to all pigments are as follows:

- ISO-787 General methods of test for pigments and extenders
- ISO-8780 Methods of dispersion for assessment of dispersion characteristics

Other ISO standards pertain to particular classes or categories of pigments, based on their chemical composition, such as ultramarine pigments, titanium dioxide, iron oxide pigments, and so forth.

Many manufacturers of paints, inks, textiles, plastics, and colours have voluntarily adopted the Colour Index International (CII) as a standard for identifying the pigments that they use in manufacturing particular colours. First published in 1925, and now published jointly on the web by the Society of Dyers and Colourists (United Kingdom) and the American Association of Textile Chemists and Colourists (USA), this index is recognized internationally as the authoritative reference on colourants. It encompasses more than 27,000 products under more than 13,000 generic colour index names.

In the CII schema, each pigment has a generic index number that identifies it chemically, regardless of proprietary and historic names. For example, Phthalo Blue has been known by a variety of generic and proprietary names since its discovery in the 1930s. In much of Europe, phthalocyanine blue is better known as Helio Blue, or by a proprietary name such as Winsor Blue. An American paint manufacturer, Grumbacher, registered an alternate spelling (Thalo Blue) as a trademark. Colour Index International resolves all these conflicting historic, generic, and proprietary names so that manufacturers and consumers can identify the pigment (or dye) used in a particular colour product. In the CII, all Phthalo Blue pigments are designated by a generic colour index number as either PB15 or PB16, short for pigment blue 15 and pigment blue 16. (The two forms of Phthalo Blue, PB15 and PB16, reflect slight variations in molecular structure that produce a slightly more greenish or reddish blue.)

Scientific and Technical Issues

Selection of a pigment for a particular application is determined by cost, and by the physical properties and attributes of the pigment itself. For example, a pigment that is used to colour glass must have very high heat stability in order to survive the manufacturing process; but, suspended in the glass vehicle, its resistance to alkali or acidic materials is not an issue. In artistic paint, heat stability is less important, while lightfastness and toxicity are greater concerns.

The following are some of the attributes of pigments that determine their suitability for particular manufacturing processes and applications:

- Lightfastness;
- Heat stability;
- Toxicity;
- Tinting strength;
- Staining;
- Dispersion;
- Opacity or transparency;
- Resistance to alkalis and acids; and
- Reactions and interactions between pigments

Swatches

Pure pigments reflect light in a very specific way that cannot be precisely duplicated by the discrete light emitters in a computer display. However, by making careful measurements of pigments, close approximations can be made. The Munsell Colour System provides a good conceptual explanation of what is missing. Munsell devised a system that provides an objective measure of colour in three dimensions: hue, value (or lightness), and chroma. Computer displays in general are unable to show the true chroma of many pigments, but the hue and lightness can be reproduced with relative accuracy. However, when the gamma of a computer display deviates from the reference value, the hue is also systematically biased.

The further a display device deviates from these standards, the less accurate these swatches will be Swatches are based on the average measurements of several lots of single-pigment watercolour paints, converted from Lab colour space to sRGB colour space for viewing on a computer display. Different brands and lots of the same pigment may vary in colour. Furthermore, pigments have inherently complex reflectance spectra that will render their colour appearance greatly different depending on the spectrum of the source illumination; a property called metamerism. Averaged measurements of pigment samples will only yield approximations of their true appearance under a specific source of illumination. Computer display systems use a technique called chromatic adaptation transforms to emulate the correlated colour temperature of illumination sources, and cannot perfectly reproduce the intricate spectral combinations originally seen. In many cases the perceived colour of a pigment falls outside of the gamut of computer displays and a method called gamut mapping is used to approximate the true appearance. Gamut mapping trades off any one of Lightness, Hue or Saturation accuracy to render the colour on screen, depending on the priority chosen in the conversion's ICC rendering intent.

Index

D

E

F

G

H

I

❑❑❑